A HISTORY IN SUM

A History in Sum

150 YEARS OF MATHEMATICS
AT HARVARD (1825–1975)

Steve Nadis and Shing-Tung Yau

Harvard University Press

Cambridge, Massachusetts

London, England

2013

Copyright © 2013 by the President and Fellows of Harvard College
All rights reserved
Printed in the United States of America

Library of Congress Cataloging-in-Publication Data

Nadis, Steven J.
A history in sum : 150 years of mathematics at Harvard (1825–1975) /
Steve Nadis and Shing-Tung Yau.
      pages cm
Includes bibliographical references and index.
ISBN 978-0-674-72500-3 (alk. paper)
1. Mathematics—Study and teaching—Massachusetts—History. 2. Harvard
University. Dept. of Mathematics. I. Yau, Shing-Tung, 1949– II. Title.
QA13.5.M43H376 2013
510.71′17444—dc23   2012049485

*To Harvard mathematicians—past, present, and future—
and to mathematicians everywhere who have
contributed to this beautiful subject.*

CONTENTS

	Preface	ix
	Prologue: The Early Days—A "Colledge" Riseth in the Cowyards	1
1	Benjamin Peirce and the Science of "Necessary Conclusions"	7
2	Osgood, Bôcher, and the Great Awakening in American Mathematics	32
3	The Dynamical Presence of George David Birkhoff	56
4	Analysis and Algebra Meet Topology: Marston Morse, Hassler Whitney, and Saunders Mac Lane	86
5	Analysis Most Complex: Lars Ahlfors Gives Function Theory a Geometric Spin	116
6	The War and Its Aftermath: Andrew Gleason, George Mackey, and an Assignation in Hilbert Space	141
7	The Europeans: Oscar Zariski, Richard Brauer, and Raoul Bott	166
	Epilogue: Numbers and Beyond	204
	Notes	211
	Index	241

PREFACE

An esteemed colleague in our department recently asked about the motivation for writing a book about the history of mathematics at Harvard. He didn't see the history of our department—or of any department, for that matter—as constituting a worthy end in itself, or at least not worthy of a book. "I don't see history as an end," he said. "I see it as a means. But if it's supposed to be an end, you really ought to explain why you consider it important and something that others, outside of this place, might actually find interesting."

I must admit to being taken aback by his remarks, as I had pretty much assumed from the get-go that the topic was, without question, meritorious. But as the person who initiated this project—at a time when I was still the department chair—I am grateful that he asked, since it forced my coauthor and me to think long and hard about the book's premise. After a good deal of reflection on this matter, I must take issue with my esteemed colleague, for in the tales of Harvard mathematicians over the years and decades, I see both a means and an end. First, there's the potential educational value in reading about the remarkable feats of remarkable people—people who at various junctures in history did indeed change the course of mathematics. Second, there are some very good stories here—stories that deserve to be told about individuals who came to mathematics through varied and singular routes, in some cases overcoming considerable adversity to pursue their respective callings.

But beyond that, I truly believe that good mathematicians (as well as good scientists in general) really need to understand their origins. By looking at the contributions of the great men and women from the past, we can trace a path showing how the important ideas in math evolved. And by looking at that path, we may gain helpful clues regarding avenues that are likely to be fruitful in the years ahead. One hope for this book, to put it another way, is that in celebrating this department's storied past, we may pave the way toward accomplishments in the future, thereby helping to ensure that its future may eventually be as storied as its past, if not more so.

You can be the most brilliant mathematician in the world, but if you try to prove a theorem without knowing anything of its history, your chances for success may be limited. It's pretty much a given that one person, no matter how tremendous a genius he or she may be, cannot go far in mathematics without taking advantage of the cumulative knowledge of those who have come before.

You might think that the theorem you just proved is the greatest thing ever—a guaranteed "game changer" and an instant classic. But in the grand scheme of things, it's merely one discrete achievement—a drop in the vast bucket we call mathematics. When you combine that "drop" with all of your other achievements, you might have produced a small volume of water—perhaps a cupful (or pitcherful)—altogether. That cup of water, as I see it, doesn't just sit in an engraved mug, alongside the degrees and awards lining our office walls. Instead, it's part of a great river that's been flowing for a long time and, I hope, will continue to flow into the indefinite future. When I do mathematics, I like to know, whenever possible, where that river has come from and where it is headed. Once I know that, I can have a better sense of what I ought to try next.

These are a few thoughts concerning the value that I see in delving into the past, in mathematics as well as in other intellectual endeavors. That still leaves the question of why we chose to write about mathematics at Harvard per se, as opposed to somewhere else, and why we consider this place significant enough to warrant such treatment. Apart from the obvious fact that I work at Harvard, and have been fortunate enough to have been employed here for the past quarter century, it is also a fact that this university has helped drive the development of mathematics in America and beyond. Until a hundred or so years ago, the field was dominated almost exclusively by Europeans. But American mathematicians have made their mark in the past century, and Harvard has been at the center of many critical advances, with our scholars continuing to play a leading role.

I won't stick my neck out and claim that Harvard is the best, which would make me unpopular in some quarters and would be hard to prove in any case. It might not even be true. But I think most objective observers would agree that the school's math department is at least one of the best. What I can say, without hesitation, is that it has produced and attracted some tremendous mathematicians, and that's been the case for more than a century. It's also an environment that has spawned some truly amazing work, and I've been struck by Harvard's illustrious tradition—and sometimes even awed by it—ever since coming here in 1987.

The rooms, libraries, and hallways of our university have been host to the exploits of legendary individuals—folks with names like Peirce, Osgood, Bôcher, Birkhoff, Morse, Whitney, Mac Lane, Ahlfors, Mackey, Gleason, Zariski, Brauer, Bott, and Tate. The influence of these scholars is still quite palpable, and their legacy is inspiring. In a half dozen or so separate fields—such as analysis, differential geometry and topology, algebraic geometry and algebraic topology, representation theory, group theory, and number theory—Harvard has led the way.

In telling the story of the pioneers in these fields, my coauthor and I aimed for something far broader than merely recounting the most notable successes to have emerged from this department. Instead, we hope we have provided a guide to a broad swath of modern mathematics, explaining concepts to nonspecialists that even mathematics students are not normally introduced to until graduate-level courses. Although lay readers will not be able to master these advanced subjects from our comparatively brief accounts, they can at least get a flavor of the work and perhaps get the gist of what it's about. As far as we know, this kind of discussion—on such abstruse topics as Stiefel-Whitney class, quasi-conformal mappings, étale cohomology, and Kleinian groups—has never before been made available to anyone but advanced students or professional mathematicians. In these pages, we hope to provide a general, and gentle, introduction to concepts that people may have heard about but have no idea what they really are.

But we also felt, from the very outset of this project, that we could not write about "important" Harvard mathematicians without explaining, at some level, what these people did that makes them important far beyond the confines of Harvard itself. Their mathematical contributions are told as part of their life stories—an approach that we hope humanizes and enlivens what might otherwise be a dry treatment of the subject.

A math department, of course, is more than just an assortment of people, lumped together through a more-or-less common academic pursuit. A department has a history, too, and its origins at Harvard were anything but grandiose. One could say that the department officially began in 1727, when the first Hollis Professor of "Mathematics and Natural Philosophy," Isaac Greenwood, was appointed. (Although the mathematics department consisted of just a single person at the time, earlier in the school's history there were no departments at all. At the school's very beginning, one instructor was responsible for teaching *all* subjects.) A man of obvious mathematical acumen, and a Harvard graduate to boot, Greenwood retained his position for eleven years, until his career at the college—and eventually his life—was done in by a weakness for alcohol.

While the department started with just one (talented but flawed) individual almost three hundred years ago, it has evolved to the point where it's now a major force in a variety of mathematical subjects, despite its relatively modest size of about two dozen junior and senior faculty members. Having grown from next to nothing to its current stature as a world leader, Harvard math can serve as a model for what it takes to build and maintain a first-class department—what it takes to be successful and productive in mathematics.

Part of the department's success, pretty much since the dawn of the twentieth century, stems from maintaining an atmosphere that encourages research among the faculty and students—even among young students, including undergraduates. Decisions over the hiring of tenured faculty are of paramount importance and can take years to be finalized, with the goal being to appoint an individual who is considered the best in a given field. Efforts are also made to keep a mix of professors, spread across different age groups, to establish balance and a continuing sense of renewal.

To the extent that the department has realized its goals and achieved distinction in various fields—and perhaps even preeminence—its history may constitute an important part of the history of mathematics in this country and, in some cases, the world. In those instances, particularly since the early twentieth century, when Harvard has played a trailblazing role, the school's math department—the people it's drawn, the avenues they've explored, and the advances they've made—has left a lasting mark on the development of mathematics everywhere. The history of our department, in other words, is a part—and I'd say a significant one—of the history of contemporary mathematics as a whole.

A department, as stated before, is more than a bunch of names listed together on a web page or catalog, or the building that these people occupy. It's like a family, with its own past, a unique genealogy, and complex dynamics—a mixture of camaraderie, goodwill, and cooperation, along with the inevitable rivalries, grudges, and power struggles. Mathematics, of course, is a broad subject, and no single person can stay on top of it all. That's one of the reasons we collaborate, interacting with people all over the country and the world. But many of our most intimate relations are with people in our own department, who may all be housed under the same roof (which, thankfully, happens to be our present circumstance)—save for the visiting scholars, scattered across the globe, who periodically come to enrich this place and expose us to fresh approaches. We learn from all of these people in important ways; they keep us abreast of developments we're not aware of and complement our knowledge and skills.

For example, George David Birkhoff (about whom much more will be said) sparked the interest of his graduate students Marston Morse and Hassler Whitney, which led them to forge novel pathways in topology. Morse, in turn, influenced Raoul Bott, who took topology into previously unforeseen areas, some of which have since given rise to key developments in both math and physics. Bott's student Stephen Smale carried this work, which built upon Morse's theory, even further still. It's all part of that "river" I referred to before. The flow is not confined to a single department, of course, but important tributaries may run through it, gaining strength (and additional "water") during the passage.

Charting the flow of this river—tracing it back to the headwaters and following the many branches at their points of confluence and divergence—is not an easy task. "Doing mathematical research is known to be hard," claimed Saunders Mac Lane, a mathematician who spent much of his early career at Harvard before moving to the University of Chicago. "Writing on the history of mathematics is not hard in the same way, but it is difficult. Part of the difficulty is that of picking the right things to bring out." History is also difficult, Mac Lane wrote, "because the connections that matter are usually numerous, often hidden, and then subsequently neglected" ("Addendum," in *A Century of Mathematics in America,* Part 3, 1989).

In keeping with the river analogy, a mathematics department, as with the field itself, is fluid rather than fixed. People are dynamic players, constantly coming and going, which means that our story is by no means limited to Cambridge, Massachusetts. Some of the mathematicians discussed here might have come to Harvard as undergraduates and returned as junior faculty, only to move on to other institutions, or they might have come for graduate school or as already-established senior faculty members. Similarly, top scholars regularly visit from other American institutions or from Europe, Asia, and elsewhere to exchange ideas with our students and faculty and to engage in research partnerships that sometimes span decades. The people based here, conversely, also travel, often collaborating with other researchers spread across the country and the globe—all of which means that our focus is far less parochial than the topic of Harvard mathematics might initially suggest. The mathematical discussion, as a result, is not restricted to the goings-on at a particular campus but is instead more international in scope.

In taking on a sprawling topic like this, one of the big challenges then is "picking the right things," as Mac Lane put it. At pivotal moments in this book, we had to make some tough decisions about whose stories would figure most prominently among many worthy contenders, while also making decisions about the time frame under consideration. Although we've tried to focus on those Harvard researchers (primarily faculty members) that made the greatest contributions to mathematics, there is, I admit, a degree of arbitrariness and subjectivity at play here. Because of limitations of time, space, and knowledge (on the part of the authors), many outstanding individuals may have been given short shrift in our account, and for that we humbly apologize.

The time frame, too, is also somewhat arbitrary. One hundred and fifty years—a century and a half—seems like a nice, round number; 1825 was singled out as the "official" starting point (though earlier years get brief mention), for that was the year in which Benjamin Peirce first came to Harvard, enrolling as a freshman at the age of sixteen. Many regard Peirce as the first American to have produced original work in the realm of pure mathematics. Within months of his appointment to the Harvard faculty in 1831, for example, Peirce proved a theorem (discussed in Chapter 1) concerning the minimum number of prime factors that an odd "perfect number"—assuming such a thing exists—must have.

Unfortunately, university officials did not reward Peirce for these efforts. They urged him to devote his energies, instead, toward the writing of textbooks, which was deemed the appropriate and, indeed, loftiest objective for a Harvard professor. In fact, little original mathematics research was being done at Harvard (or at any other American university, for that matter) until the late nineteenth and early twentieth centuries. This transition—the coming of age of mathematics at Harvard in concert with parallel developments elsewhere in the country—is the subject of Chapter 2. It was also taken up in an excellent 2009 article by Steve Batterson, "Bôcher, Osgood, and the Ascendance of American Mathematics at Harvard," which was published in the *Notices of the American Mathematical Society*. While I found Batterson's account fascinating, I felt that it ended just as the story was getting interesting—just when our department was starting to hit its stride. That, indeed, was part of the motivation for this book—to write about what happened once mathematics really took hold at Harvard at the turn of the twentieth century.

As I see it, a tradition of excellence has been established here that is self-perpetuating, having taken on a life of its own. The story is constantly unfolding, with faculty members, research fellows, graduate students, and undergraduates continuing to do impressive research, proving new theorems, and winning prestigious awards. Since there's no obvious cutoff point to this work, we made the decision (again somewhat arbitrarily) to essentially cap our chronicle in the year 1975 or thereabouts—the rationale being that it takes some time, perhaps a matter of decades, to accurately appraise developments in mathematics. There are many theorems that people initially get excited about, but then twenty to thirty years later we find that some of them do not loom quite so large after all.

One consequence of our decision regarding the time frame is that, with a few exceptions, we are writing about people who are no longer in the department and most of whom are deceased. That makes it easier when drafting a history like this, since it's hard to identify a person's most salient achievements while his or her career is still in midstream. It's helpful to have the benefit of time in assessing the weight of one's accomplishments. And there's always the chance that, at any given moment, the next thing that he or she does may eclipse everything that preceded it.

The downside of this strategy is that we inevitably omit a lot of extraordinary mathematics, because it's clear that Harvard scholars have had many successes in the years since 1975. Perhaps, someday, there will be a sequel to this narrative in which we read about their stories and accomplishments as well.

—Shing-Tung Yau

When first approached by my coauthor to take on this project, I must confess that I didn't know what I was getting into. (That, I'm embarrassed to admit, is the case with most of the literary endeavors I get involved with.) Although I'd been in the math department on countless occasions before—having met many faculty members, students, and postdocs during those visits—I'd never given much thought, if any, to the setting or context in which these people worked. I had no sense of how they fit into the bigger fabric here. Popping in and out, as I often did, one can easily overlook the fact that this place is steeped in tradition. Upon a bit of digging, however, I was pleased to find a rousing cast of characters, over the decades and centuries, who'd done so much—far more than I'd realized—to further the cause of mathematics in this country and throughout the world. I was eager to learn more about them and what they had achieved, and I hoped that others—who, like me, had no formal connection with this place—would find their stories engaging as well.

A mathematician I spoke with, the editor of a prominent mathematics journal, told me that Harvard was special—"a beacon in mathematics," as he put it. "Almost every mathematician who comes to the U.S. from afar wants to stop at Harvard sometime during his or her visit." Before embarking on this project, I'd never heard anyone make a statement like that, and the fact that someone unaffiliated with Harvard would say that is certainly a tribute to the department. But it's also true that Harvard mathematics, regardless of its present standing, was not always a beacon. For a long time, Harvard mathematicians, as well as American mathematicians in general, were not making lasting contributions to their field. That has changed, of course, which is why my coauthor and I considered writing a book on this subject. We thought it might be instructive to see how the department rose from humble beginnings—like most of its American counterparts—to its current position of prominence. Our focus is not on the evolution of course curricula, innovations in math education, or shifts in administrative policies but, rather, on noteworthy *achievements* in mathematics—spectacular results, made by fascinating people, that have stood the test of time.

That quest has involved a fair amount of research, interviews, and general investigation—for which we have relied on the help of a large number of people, both inside and outside the department. We'd now like to thank as many of them as we can, apologizing to anyone whose efforts were overlooked amidst this frantic activity, which (as the word "frantic" implies) did not always proceed in the most orderly fashion. Thanks are owed to Michael Artin, Michael Atiyah, Michael Barr, Ethan Bolker, Joe Buhler, Paul Chernoff, Chen-Yu Chi, John Coates, Charles Curtis, David Drasin, Clifford Earle, Noam Elkies, Carl Erickson, John Franks, David Gieseker, Owen Gingerich, Daniel Goroff, Fan Chung Graham, Robert Greene, Benedict Gross, Michael Harris, Dennis Hejhal, Aimo Hinkkanen, Eriko Hironaka, Heisuke Hironaka, Roger Howe, Yi Hu, Norden Huang, Lizhen Ji, Yunping Jiang, Irwin Kra, Steve Krantz, Bill Lawvere, Peter Lax, Jun Li, Bong Lian, David Lieberman, Albert Marden, Brian Marsden, Barry Mazur, Colin McLarty, Calvin Moore, Dan Mostow, David Mumford, Richard Palais, Wilfried Schmid, Caroline Series, Joseph Silverman, Robert Smith, Joel Smoller, Shlomo Sternberg, Dennis Sullivan, Terence Tao, John Tate, Richard Taylor, Andrey Todorov, Howell Tong, Henry Tye, V. S. Varadarajan, Craig Waff, Hung-Hsi Wu, Deane Yang, Lo Yang, Horng-Tzer Yau, Lai-Sang Young, and Xin Zhouping. In particular, Antti Knowles, Jacob Lurie, and Loring Tu were extremely generous with their time, and the authors are grateful for their invaluable input. Maureen Armstrong, Lily Chan, Susan Gilbert, Susan Lively, Rima Markarian, Roberta Miller, and Irene Minder provided vital administrative assistance. The librarians at the Harvard University Archives have been extremely helpful, as was Nancy Miller of Harvard's Birkhoff Library and Gail Oskin and others at Harvard Photographic Services. We'd also like to thank our editor, Michael Fisher, and his colleagues at Harvard University Press—including Lauren Esdaile, Tim Jones, Karen Peláez, and Stephanie Vyce—for taking on this project and converting our electronic files into such a handsome volume. Brian Ostrander and the folks at Westchester Publishing Services, along with copy editor Patricia J. Watson, helped put the finishing touches on our book; we appreciate the services as well as the closure.

The authors benefitted from the kind support of Gerald Chan, Ronnie Chan, and the Morningside Foundation, without which we would not have been able to complete this project. We owe them a debt of gratitude and will not forget their generosity.

Finally, we'd like to pay tribute to our families, who always have to put up with a lot when one member of the clan decides to abandon reason and get involved in something as all-consuming as writing a book. My coauthor thanks his wife, Yu-Yun, and his sons, Isaac and Michael; I thank my wife, Melissa, my daughters, Juliet and Pauline, and my parents, Lorraine and Marty, for their support and unwavering patience.

They heard more about Harvard mathematics than the average person will be exposed to, never once hinting that the contents of that discussion were anything less than riveting.

—Steve Nadis

Mathematics is the science which draws necessary conclusions.
—BENJAMIN PEIRCE, 1870

PROLOGUE

The Early Days—A "Colledge" Riseth in the Cowyards

The beginnings of Harvard University (originally called "Harvard Colledge" in the vernacular of the day) were certainly humble, betraying little hints of what was in store in the years, and centuries, to come. The school was established in 1636 by decree of the Great and General Court of the Massachusetts Bay Colony, but in that year it was more of an abstraction than an actual institute of higher learning, consisting of neither a building nor an instructor and not a single student. In 1637, or thereabouts, a house and tiny parcel of cow pasture were purchased in "Newetowne" (soon to be renamed Cambridge) from Goodman Peyntree, who had resolved to move to Connecticut, which was evidently the fashionable thing to do at the time among his more prosperous neighbors. In that same year, the college's first master was hired—Nathaniel Eaton, who had been educated at the University of Franeker in the Netherlands, where he had written a dissertation on the perennially enthralling topic of the Sabbath. At first, it was just Eaton, nine students, and a farmhouse on little more than an acre of land. John Harvard, a minister in nearby Charlestown, who was a friend of Eaton's and "a godly gentleman and lover of learning,"[1] died in 1638, having bequeathed the fledgling school half of his estate and his entire four-hundred-volume library.

Some 375 years later, the university that bears John Harvard's name still stands on that former cow patch—albeit with some added real estate—the oldest institution of higher learning in the United States. The school's libraries collectively hold more than sixteen million books, compared with the few hundred titles in the original collection. The number of students has similarly grown from a handful to the more than 30,000 that are presently enrolled on a full- or part-time basis. In place of the lone schoolmaster of the 1630s, there are now about 9,000 faculty members (including those with appointments at Harvard-affiliated teaching hospitals), plus an additional 12,000 or so employees. Eight U.S. presidents have graduated from the university, and its faculty has produced more than forty Nobel laureates. Harvard's professors and its graduates have won seven Fields Medals—sometimes called the mathematics equivalent of a Nobel Prize—and account for more than one-quarter of the sixty-two presidents of the American Mathematical Society. In addition, Harvard scholars have earned many other prestigious mathematics honors about which more will be said in the pages to come.

None of this, of course, could have been foretold when the school was started by Puritan colonists, who—in contrast to today's liberal arts philosophy—were fearful, perhaps above all else, of leaving "an illiterate Ministry to the Churches, when our present Ministers shall lie in the Dust."[2] The founders felt a pressing need to train new ministers and to produce a citizenry capable of reading the Bible and hymnbooks—among other literature—placed before them.

What the founders had in mind, in other words, was something along the lines of a glorified Bible study school, and the "colledge" they launched for this purpose surely got off to a rocky start. The school's first hire, Eaton, had a tendency to "drive home lessons with the rod." In 1639, the second year of his tenure, he beat his assistant with "a walnut-tree cudgel big enough to have killed a horse," and the assistant might have died had it not been for the timely intervention of a minister from the church nearby. Eaton was hauled into court for assault and dismissed from his position in that same year—partly owing to his fondness for corporal punishment and partly owing to his wife's substandard cooking, which left the students ill fed and ill tempered. Evidently, she offered them too little beef (or none at all) and bread "sometimes made of heated, sour meal," and—perhaps the gravest offense of all—she sometimes made the boarders wait a week between servings of beer. In the absence of any headmaster or teacher of any sort, the school closed its doors during the 1639–40 academic year, and students were sent elsewhere—some back to the farms whence they came—prompting many to wonder whether the school would ever reopen.[3]

Harvard's overseers had better luck with Eaton's successor, a University of Cambridge graduate named Henry Dunster, who put the school on a sounder course, both fiscally and academically, during his fourteen years as master and president. Dunster devised a three-year, three-pronged educational plan that revolved around liberal arts, philosophies, and languages (or "learned tongues," as they were called). Dunster's program remained largely intact long after he resigned in 1654, extending well into the eighteenth century.

While the curriculum offered students a reasonably broad foundation, it was, in the words of the historian Samuel Eliot Morison, "distinctly weak" in mathematics and the natural sciences—in keeping with the example of English universities of the era upon which Harvard was modeled.[4] ("The fountain could not rise higher than its source," another historian once explained, in reference to the paucity of mathematics instruction to be found on campus.)[5] Since "arithmetic and geometry were looked upon . . . as subjects fit for mechanics rather than men of learning," Morison adds,[6] exposure to these subjects was limited to pupils in the first three quarters of their third and final year of study, with the fourth quarter of that year reserved for astronomy. Students met at 10 A.M. on Mondays and Tuesdays for the privilege of honing their mathematical skills. These times were apparently etched in stone, or etched into the school's bylaws, which stated that the hours were not subject to change "unless experience shall show cause to alter."[7]

In the first one hundred or so years, mathematics instructors, who held the title of tutors, had little formal training in the subject—consistent with the general sentiment that the subject itself hardly warranted a more serious investment. Students, similarly, had to demonstrate proficiency in Latin ("sufficient to understand Tully, or any like classical author") to gain admittance to Harvard but faced no entrance examinations in mathematics and "were required to know not even the multiplication table."[8]

Evidence suggests there was little change in mathematics education at Harvard for another eighty to ninety years after Dunster introduced his original course of study. "Arithmetic and a little geometry and astronomy constituted the sum total of the college instruction in the exact sciences," wrote Florian Cajori in an 1890 review of mathematics training in this country. "Applicants for the master's degree only had to go over the same ground more thoroughly."[9]

Algebra, for example, probably did not show up in the Harvard curriculum until the 1720s or 1730s, Cajori contended, even though the French mathematician and philosopher René Descartes introduced modern algebraic notation in 1637. A textbook on the subject, *Elements of That Mathematical Art Commonly Called Algebra* by an English schoolteacher, John Kersey, was published in two volumes in 1673 and 1674, nearly a half century before Harvard saw fit to expose its students to algebra.

Based on senior thesis titles of the day, the mathematics scholarship that took place was hardly earth-shattering, Morison writes, "consisting largely of such obvious propositions as: 'Prime numbers are indivisible by any factor' and 'In any triangle the greater side subtends the greater angle.'"[10] It seems evident that no new earth was being tilled, nor new treasures dug up, in this agrarian milieu.

A turning point came in 1726, when the first mathematics professor, Isaac Greenwood, was appointed. Greenwood, a Harvard graduate, did much to raise the level of pedagogy in science and math, offering private lessons on various advanced topics. He also gave a series of lectures and demonstrations on the discoveries of Isaac Newton, who coincidentally died in the same year, 1727, that Greenwood became the first occupant of a newly endowed chair, the Hollis Professorship of Mathematics and Natural Philosophy—named after Thomas Hollis, a wealthy London-based merchant and Harvard benefactor. Greenwood was responsible for many other firsts, as well, authoring the first mathematics text written in English by a native-born American and being the first mathematics professor to teach calculus in the colonies. He also taught algebra and was possibly the first to introduce the subject to Harvard students.

Despite these virtues, Greenwood let his taste for alcohol get the better of him. After repeated bouts of drunkenness and failures to abstain from liquor, despite being granted many opportunities to mend his ways, he was permanently discharged from his position in 1738, fired for "gross intemperance."[11] His dismissal was described in an early history of the university as an "excision of a diseased limb from the venerable trunk of Harvard."[12] Greenwood became a traveling lecturer after leaving Harvard and, sadly, drank himself to death seven years later.

John Winthrop, Greenwood's twenty-four-year-old successor, fared considerably better, holding the Hollis professorship for forty-one years. He was "the first important scientist or productive scholar on the teaching staff at Harvard College," according to Morison, who compared Winthrop with Benjamin Franklin in terms of versatility: "With the time and means at his disposal, he was able to carry investigation deeper than Franklin on many subjects." Winthrop studied electricity, sunspots, and seismology, "proving that earthquakes were purely natural phenomena, and not manifestations of divine wrath," thereby incurring the (undivine) wrath of some clergymen.[13]

Although Winthrop was a first-rate scientist and, by all accounts, an excellent teacher, Julian Coolidge (a member of Harvard's math faculty from 1899 to 1940) could not say "that his interest in pure mathematics was outstanding"—perhaps a symptom of the times.[14] As a general rule, Cajori noted, "the study of pure mathematics met with no appreciation and encouragement. Original work in abstract mathematics would have been looked upon as useless speculations of idle dreamers."[15]

The next two occupants of the Hollis mathematics professorship, Samuel Williams and Samuel Webber, were less distinguished, according to Coolidge, who claimed "there was certainly a retrocession in ... interest in mathematics during these years."[16] Williams, who conducted research on astronomy, meteorology, and magnetism, was an active socialite with an extravagant lifestyle that put him in serious debt—and ultimately out of his Harvard job in 1788. Webber, described as "a man without friends or enemies," assumed the Hollis chair in 1789, becoming president of the college in 1806, though Morison characterized him as "perhaps the most colorless President in our history." He died in 1810, long before his dreams of establishing an astronomical observatory at Harvard were realized, with his only tangible accomplishment on that front being rather modest: the construction of an "erect, declining sundial."[17]

In 1806, the Hollis chair was offered to Nathaniel Bowditch, a self-taught mathematician of growing repute, who turned down the offer to pursue other interests. A year later, the mathematics and natural philosophy chair was filled by John Farrar, a scientist and Harvard graduate who would later transform our conception of hurricanes, writing that the great gale that struck New England in 1815 "appears to have been a moving vortex and not the rushing forward of a great body of the atmosphere."[18] Although Farrar did not complete any original mathematics research of note, he was an inspired lecturer who brought modern mathematics into the Harvard curriculum, personally translating the works of French mathematicians such as Jean-Baptiste Biot, Étienne Bézout, Sylvestre Lacroix, and Adrien-Marie Legendre.

Harvard undergraduates began studying Farrar's formulation of Bézout's calculus in 1824. A year later, a precocious freshman named

Benjamin Peirce, who had already studied mathematics with Bowditch, enrolled in the school. His father, also named Benjamin Peirce, was the university librarian who would soon write the history of Harvard.[19] His son, meanwhile, would soon rewrite the history of mathematics—both at Harvard and beyond.

1

BENJAMIN PEIRCE AND THE SCIENCE OF "NECESSARY CONCLUSIONS"

Benjamin Peirce came to Harvard at the age of sixteen and essentially never left, all the while clinging to the heretical notion that mathematicians ought to do *original* mathematics, which is to say, they should prove new theorems and solve problems that have never been solved before. That attitude, sadly, was not part of the orthodoxy at Harvard, nor was it embraced at practically any institution of higher learning in the United States. At Harvard and elsewhere, the emphasis was on *teaching* math and *learning* math but not on *doing* math. This approach never sat well with Peirce, who was unable, or unwilling, to be just a passive recipient of mathematical doctrine. He felt, and rightfully so, that he had something more to contribute to the field than just being a good reader and expositor. Consequently, he was driven to advance mathematical knowledge and disseminate his findings, even though the university he worked for did not share his enthusiasm for research or mathematics journals. (The "publish or perish" ethic, evidently, had not yet taken hold.)

When Peirce was just twenty-three years old, newly installed as a tutor at Harvard, he published a proof about perfect numbers: positive integers that are equal to the sum of all of their factors, including 1. (Six, for instance, is a perfect number: its factors, 3, 2, and 1, add up to 6. Twenty-eight is another example: $28 = 14 + 7 + 4 + 2 + 1$.) All the perfect numbers known at that time—and still to this day—were even. Peirce wondered whether odd perfect numbers might exist, and his proof, which is discussed later in this chapter, placed some constraints on their existence. Despite the fact that this work turned out to be more than fifty years ahead of its time, it did not garner international acclaim—or any notice, for that matter—mainly because the leading European scholars did not take American mathematics journals seriously, nor did they expect them to publish anything of note. Nevertheless, Peirce's accomplishment did signal, to anyone who might have been paying attention, that a new era of mathematics was starting at Harvard—one that the school's administration could not suppress, even though it did nothing to encourage Peirce in this direction.

Peirce had, however, received strong encouragement from Nathaniel Bowditch, who was considered one of the preeminent mathematicians in the United States. Bowditch helped cultivate Peirce's interest in "real," cutting-edge mathematics, and had Bowditch made a different career decision, he might have played an even more direct role in his protégée's education. In 1806, Harvard offered Bowditch the prestigious Hollis Chair of Mathematics and Natural Philosophy. Bowditch turned down that offer, just as he turned down subsequent offers from West Point and the University of Virginia. But he did not entirely turn his back on Harvard; he later served as a fellow to the Harvard Corporation during a term that overlapped with Peirce's years there as a student, tutor, and faculty member.

As leading mathematicians go, Bowditch was something of an anomaly. He was almost entirely self-educated; he had never gone to college, nor did he attend high school. Instead, he left school at the age of ten to join the workforce, assisting his father in the cooper trade, making barrels, casks, and other wooden vessels. He helped his father for two years and then joined the shipping industry. After voyaging to distant places like Sumatra and the Philippines, he returned to Massachusetts where he entered the insurance business, while resuming his mathematical studies on the side. Although his exposure to formal education was brief, he had learned enough math on his own to know that a university could never offer him as much money as he came to earn in his job as president of the Essex Fire and Marine Insurance Company.

Bowditch nevertheless continued to pursue his interest in mathematics, focusing on celestial mechanics—the branch of astronomy that involves the motions of stars, planets, and other celestial objects. By 1806, the year Bowditch was recruited by Harvard, he had read all four volumes of Pierre-Simon Laplace's treatise *Mécanique Céleste*. (The fifth volume came out in 1825.) Bowditch, in fact, did a good deal more than just read it; he set about the task of translating the first four volumes of Laplace's great work. His efforts went beyond mere translation—no mean task in itself—and included a detailed commentary that helped bring Laplace within the grasp of American astronomers and mathematicians, who, for the most part, had not been able to understand his treatise before. Bowditch not only brought Laplace's work up to date but also filled in many steps that the original author had omitted. "I never came across one of Laplace's 'thus it plainly appears' without feeling sure that I have hours of hard work before me to fill up the chasm and find out and show how it plainly appears," Bowditch said.[1] The French mathematician Adrien-Marie Legendre praised Bowditch's efforts: "Your work is not merely a translation with a commentary; I regard it as a new edition, augmented and improved, and such a one as might have come from the hands of the author himself if he had consulted his true interest, that is, if he had been solicitously studious of being clear."[2]

Peirce, who was born in Salem, Massachusetts, in 1809, would probably have met Bowditch eventually, given Peirce's manifest talent in mathematics and Bowditch's growing reputation in the field. But they met earlier than they might have otherwise because Peirce went to a grammar school in Salem where he was a classmate and friend of Henry Ingersoll Bowditch, Nathaniel's son. The story has it that Henry showed Peirce a mathematical problem that his father had been working on. Peirce uncovered an error, which the son brought to his father's attention. "Bring me the boy who corrects my mathematics," Bowditch reportedly said, and their relationship blossomed from there.[3]

Bowditch moved from Salem to Boston in 1823. Two years later, the sixteen-year-old Peirce moved to nearby Cambridge to enter Harvard, following in the footsteps of his father, Benjamin Peirce Sr., who attended the college and later worked as the school librarian and historian. By the time the younger Peirce arrived on campus, he already had a mentor—not some street-smart upperclassman, but Bowditch himself, who was then a nationally known figure. Hard at work on his Laplace translation at the time, Bowditch enlisted the keen eye and proofreading services of the young Peirce. The improvements suggested by Peirce were reportedly "numerous."[4] The first volume of Bowditch's translation was published in 1829, the year that Peirce graduated from Harvard. The other three volumes were published in 1832, 1834, and 1839, respectively. (Independently, a separate translation of Laplace's work came out in 1831. That book, titled *The Mechanism of the Heavens,* was written by Mary Somerville, a British woman who, like Bowditch, had mostly taught herself mathematics and endeavored to make Laplace accessible.

Her book, too, went beyond a mere translation, containing detailed explanations that put his treatise into more familiar language.)[5]

Peirce continued to review Bowditch's manuscripts during his tenure as a Harvard professor. "Whenever one hundred and twenty pages were printed, Dr. Bowditch had them bound in a pamphlet form and sent them to Professor Peirce, who, in this manner, read the work for the first time," wrote Nathaniel Ingersoll Bowditch, another of Nathaniel Bowditch's sons, in a memoir about his father. "He returned the pages with the list of errata, which were then corrected with a pen or otherwise in every copy of the whole edition."[6]

In this way, Peirce was exposed from an early age to mathematics more advanced than could be found in any American curriculum—writings that other undergraduates simply were not privy to. Scholars have speculated that the excitement of reading and mastering Laplace's work may have drawn Peirce to mathematical research. It is evident that Laplace's writings made a deep impression on him. Decades later, in the pre-Civil War era, a student told Peirce that he risked incarceration for helping to rescue a runaway slave; the only consolation about being locked up in prison, the student said, was that he would finally have time to read Laplace's magnum opus. "In that case, I sincerely wish you may be," Peirce quipped.[7]

Peirce had, of course, an even deeper reverence for his mentor than he did for Laplace. Bowditch, in turn, was convinced that his young charge would go far, claiming that, as an undergraduate, Peirce already knew more mathematics than John Farrar, who then held the Hollis professorship.[8] Peirce returned the favor decades later, calling Bowditch the "father of American geometry" in a treatise he wrote on analytical mechanics that was dedicated to his mentor.[9] Before long, a similar term, "father of American mathematics," was applied to Peirce (by the British mathematician Arthur Cayley, among others). Through the force of his personality and the originality of his work, Peirce came to be known as the leading American mathematician of his generation and, more generally, as the initiator of mathematical research at American universities.[10]

On that score, Peirce faced little competition. Before he entered the scene, no one thought that "mathematical research was one of the things for which a mathematical department existed," Harvard mathematician Julian Coolidge wrote in 1924. It was certainly not a job prerequisite since there were not nearly as many people qualified to conduct high-level research, or inclined to do so, as there were available teaching slots. "Today it is commonplace in all the leading universities," Coolidge added. "Peirce stood alone—a mountain peak whose absolute height might be hard to measure, but which towered above all the surrounding country."[11]

Despite the abilities Peirce exhibited at an early age, it was not obvious that he would have the opportunity to attain the aforementioned heights. After receiving his bachelor's degree from Harvard in 1829, Peirce had essentially no options for advanced studies of mathematics in the United States, because no Ph.D. programs in math existed at the time. One could go to Europe—Göttingen, Germany, was a popular destination for mathematically inclined young Americans—but this was not a realistic possibility for Peirce, mainly for financial reasons. It appears that his family could not afford the luxury of sending him to school abroad; instead, he had to start earning a living soon after graduation.

He taught for two years at Round Hill School, a preparatory school in Northampton, Massachusetts, before returning to Harvard in 1831 to work as a tutor. But with Farrar, the Hollis chair, away in Europe at the time, Peirce was immediately placed at the head of the department. For health reasons, Farrar never resumed his full duties. Peirce continued to run the department, first as University Professor of Mathematics and Natural Philosophy, starting in 1833, and later as the Perkins Professor of Mathematics and Astronomy, starting in 1842. He retained the Perkins chair until he died in 1880—almost fifty years after joining the Harvard faculty.

Within months of his original appointment, Peirce submitted his aforementioned proof on perfect numbers to the *New York Mathematical Diary,* one of many journals to which he contributed, whereby he had gained a growing reputation as a talent to be reckoned with.[12] Peirce took the position that people needed to solve actual mathematical problems in order to earn the title of mathematician. "We are too prone to consider the mere reader of mathematics as a mathematician, whereas he does not much more deserve the name than the reader of poetry deserves that of poet," wrote Peirce, by way of promoting *Mathematical Miscellany,* a journal that he contributed to frequently and of which he eventually (though briefly) became editor.[13]

His 1832 paper on perfect numbers concerned a topic that had attracted attention since antiquity. Euclid proved in the *Elements,* which he wrote around 300 B.C., that if $2^n - 1$ is a prime number, then $2^{n-1}(2^n - 1)$

is a perfect number. Roughly 2,000 years later, Leonhard Euler proved that every even perfect number must be of this form. "But I have never seen it satisfactorily demonstrated that this form includes all perfect numbers," Peirce wrote.[14] He was alluding to the question of whether odd perfect numbers might exist. This was among the oldest open problems in mathematics, and it remains unsolved to this day. But Peirce gave a partial answer to that question, proving that an odd perfect number—if there is one—must have at least four distinct prime factors. A perfect number with fewer than four prime factors (such as 6) has to be even.

In achieving this result, modest though it may seem, Peirce was far ahead of his contemporaries. The British mathematician James J. Sylvester, who happened to be a good friend of Peirce's, and the French mathematician Cl. Servais proved the exact same thing—that any odd number must have at least four distinct prime factors—in 1888, *fifty-six years* after Peirce had established that very fact.[15] Sylvester and Servais clearly had not seen Peirce's paper, which was published in the *Mathematical Diary*—a journal that was not widely read in the United States, let alone followed by many readers outside the country. Peirce would run into this problem again and again, as he had set up shop in what was regarded by many Europeans as a mathematical backwater of the highest rank.

Later, in 1888, Sylvester proved that an odd perfect number must have at least five distinct prime factors and subsequently conjectured that there must be at least six. As of this writing, more than a century later, the minimum number of distinct prime factors now stands at nine.[16] If nothing else, Peirce started a cottage industry that persists to this day. And even after all this time, no one yet knows whether odd perfect numbers exist. But odd numbers up through 10^{300} have already been checked without success, making the prospect of finding an odd perfect number seem increasingly dim.

Curiously, Peirce's employers did not appreciate his accomplishment: proving a new theorem in number theory that related to a legendary problem. Harvard president Josiah Quincy pushed Peirce in a more conventional direction: writing textbooks. Peirce, however, had greater ambitions than that, asking whether the Harvard Corporation wanted him to "undertake a task that must engross so much time and is so elementary in its nature and so unworthy of one that aspires to anything higher in science." But the corporation agreed with Quincy's directive. The notion of doing original research in mathematics was so novel then as to have been practically unheard of in the United States; hardly anyone was qualified to even attempt it, which is why Peirce's entreaties fell on deaf ears. Instead, he published seven textbooks over the next ten years, on such subjects as plane trigonometry, spherical trigonometry, sound, plane and solid geometry, algebra, and *An Elementary Treatise on Curves, Functions, and Forces* (in two volumes). His text on analytical mechanics came much later, in 1855. And in keeping with the line taken by the Harvard administration, he published no further papers on number theory, though he did not stop pursuing original work in mathematics and science.[17]

While his textbooks were original in presentation and mathematically elegant, they were too concise for most students—largely stripped bare of exposition—which made them difficult to understand. Simply put, the texts were too demanding for all but the most exceptional students. They were "so full of novelties," explained former Harvard president Thomas Hill, "that they never became widely popular, except, perhaps, the trigonometry; but they have had a permanent influence upon mathematical teaching in this country; most of their novelties have now become commonplaces in all textbooks."[18]

Peirce's 1855 treatment of analytical mechanics, for example, did attract some favorable notice. Soon after its publication, an American student in Germany asked an eminent German professor what book he should read on that subject. The professor replied, "There is nothing fresher and nothing more valuable than your own Peirce's quarto."[19]

Despite such praise from those well versed in mathematics, the works were generally unpopular among students, some of whom wrote in their books: "He who steals my Peirce steals trash."[20] In fact, enough students complained about the impenetrability of Peirce's texts that they were investigated by the Harvard Committee for Examination in Mathematics, which concluded that "the textbooks were abstract and difficult, that few could comprehend them without much explanation, that Peirce's works were symmetrical and elegant, and could be perused with pleasure by the adult mind, but that books for young students should be more simple." The report, in other words, was all over the map, but in the end Peirce's textbooks continued to be used in Harvard classrooms for many more years.[21]

Peirce's lectures were a mixed bag as well. The average person found them almost impossible to follow—some saying that the speed of Peirce's mental processes made it difficult for him to put things in a way that others could comprehend. "In his explanations, he would take giant strides," said Dr. A. P. Peabody, a tutor during the 1832–33 academic year. "And his frequent 'you see' indicated what he saw clearly, but that of which his pupils could hardly get a glimpse."[22]

For advanced students who could keep pace with Peirce's rapid train of thought, the talks could be inspiring. "Although we rarely could follow him, we sat up and took notice," said William Elwood Byerly, Peirce's student both in college and in graduate school. Byerly earned the first Ph.D. granted by Harvard, becoming an assistant professor at the university in 1876.[23]

A Cambridge woman had a similar experience when she attended one of Peirce's lectures. "I could not understand much that he said; but it was splendid," she reported. "The only thing I now remember in the whole lecture is this—'Incline the mind to an angle of 45 degrees, and periodicity becomes non-periodicity, and the ideal becomes real.'"[24]

While Ralph Waldo Emerson once asserted that "to be great is to be misunderstood," Peirce's example at Harvard offered a variant on that dictum: to be great is to be incomprehensible. At a presentation before the National Academy of Sciences, Peirce once spent an hour filling a blackboard with dense equations. Upon turning to see the perplexed faces among the attendees, he said, "There is only one member of the Academy who can understand my work, and he is in South America."[25] Coolidge regarded Peirce as a rousing, if opaque, lecturer: "His great mathematical talent and originality of thought, combined with a total inability to put anything clearly, produced among his contemporaries a feeling of awe that amounted almost to dread."[26]

Not only were the lectures difficult to follow, Byerly noted, but also they were often ill prepared. "The work with which he rapidly covered the blackboard was very illegible, marred with frequent erasures, and not infrequent mistakes (he worked too fast for accuracy). When the college bell announced the close of the hour . . . , we filed out, leaving him abstractedly staring at his work, still with chalk and eraser in his hands, entirely oblivious of his departing class."[27]

Despite the ostensible drawbacks to Peirce's pedagogy—the words "wretched"[28] and "lamentable"[29] have been applied—former Harvard president Abbott Lawrence Lowell said that in his fifty-year association with the college, "Benjamin Peirce still impresses me as the most massive intellect with which I have ever come into close contact, and as being the most profoundly inspiring teacher that I ever had."[30] Yet Lowell did admit that Peirce's blackboard presentations left something to be desired: "He was impatient of detail, and sometimes the result would not come out right; but instead of going over his work to find the error, he would rub it out, saying that he had made a mistake in a sign somewhere, and that we should find it when we went over our notes."[31]

His "boardside" (as opposed to bedside) manner was also suspect, according to Oliver Wendell Holmes, who was a college classmate of Peirce's as well as a fellow faculty member. "If a question interested him, he would praise the questioner, and answer it in a way, giving his own interpretation to the question," Holmes said. "If he did not like the form of the student's question, or the manner in which it was asked, he would not answer it at all."[32]

In an anecdote of this sort recounted in the *Harvard Crimson,* a student took Peirce up on his offer to answer questions after class, in the event that any of his explanations on higher mathematics were not crystal clear. But after raising his question, the student received no response. He repeated the question and still received no response. "'But did you not invite us to ask you questions in regard to your lecture, sir?' inquired the student. 'Oh, certainly,' replied Professor Peirce, with an air of surprise, 'but I meant intelligent questions.'"[33]

Students attending the college from 1860 on had some relief, because those who found Peirce's soliloquies lacking in clarity could get help from his oldest son, James Mills Peirce, who filled in for his father that year and became an assistant professor a year later. (Another of his four sons, Charles Sanders Peirce, ultimately became far more famous than James, with achievements rivaling—if not exceeding—those of his father.) As Benjamin Peirce lightened his course load late in his career, James took over more and more of the teaching responsibilities in the department, ultimately taking over the Perkins chair after his father's death. The quality of instruction improved under the leadership of James, who was "a much better teacher even though he lacked the spark of originality" and made "negligible" contributions to the field of mathematics. But his contributions to the Harvard department were deeply felt, Coolidge wrote.[34]

Long before relief for students arrived in the form of his son James, the senior Peirce exhibited little patience for the mathematically dim. He preferred to spend his time with more talented pupils. Unfortunately, individuals of that sort tended to be scarce at Harvard, as well as at other American universities of the time. In 1835, Peirce proposed that students should not have to take mathematics beyond their first year unless they so chose. The university adopted his plan in 1838. "This allowed Peirce to teach more advanced mathematics than was being taught elsewhere in the United States," wrote Peirce biographer Edward Hogan.[35]

Peirce believed, further, that professors should devote more time to research and less to teaching, spending no more than two hours a day on teaching so as to have more time for original investigations. Years later, Peirce found a strong ally in Harvard president Thomas Hill—a former student who, according to historian Samuel Eliot Morison, "was said to have been the only undergraduate of his generation to comprehend" Peirce's higher mathematical demonstrations.[36] Like Peirce, Hill believed that "our best Professors are so much confined with the onerous duties of teaching and preparing lectures that they have no time nor strength for private study and the advancement of science and learning." The system's failing was especially pronounced in mathematics education, Hill said, owing to the "inverted method" adopted in so many schools of "exercising the memory, loading it with details . . . , but not illuminating the imagination with principles to guide its flight."[37] Peirce, of course, concurred wholeheartedly with Hill's assessment: "I cannot believe it to be injudicious to reduce the time which the instructor is to devote to his formal teaching to a couple of hours each day, or even less."[38]

One area in which Peirce spent considerable time outside of class was astronomy, which was not uncommon for mathematicians in that era. Both Laplace and Bowditch, as mentioned, put much effort into that area. Another contemporary, Carl Friedrich Gauss, widely regarded as one of the greatest mathematicians of all time, spent nearly the last fifty years of his life as a professor of astronomy at the Göttingen Observatory. Peirce himself played a pivotal role in the founding of the Harvard College Observatory in 1839, although it did not become a fully functioning observatory until 1847, when the first telescope, the fifteen-inch "great refractor," was installed. "This was the first great and efficient observatory to be established in the United States," wrote T. J. J. See, an astronomer at the University of Chicago. "And its value to American science may be judged from the fact that only a few years before Dr. Bowditch had declared: 'America has as yet no observatory worthy of mention.'"[39]

In the years since the observatory's founding, astronomy took up an expanding portion of Peirce's time and attention. In fact, many of his contemporaries thought of him first and foremost as an astronomer. Peirce took advantage, for instance, of the "Great Comet of 1843" (formally known as C/1843 D1 and 1843 I), which was visible in midday, to give a series of public lectures aimed at sparking public interest in astronomy. At the same time, Peirce embarked on elaborate calculations regarding the comet's orbit. This exercise would prove handy when Peirce engaged in even more involved calculations concerning the orbit of the newly discovered planet Neptune, a high-profile and contentious matter.

The story burst to the fore in 1846, when Johann Gottfried Galle of the Berlin Observatory pointed his telescope to a predetermined spot in the sky and discovered Neptune, the eighth planet from the sun. Prior to Galle's observations, two mathematician-astronomers, Urbain Jean Joseph Le Verrier of France and John Couch Adams of England, had both predicted the position in the sky of a more distant, and as yet unknown, planet in the solar system that was responsible for perturbations in the orbit of Uranus. Of the two, Le Verrier was fortunate in having access to an astronomer, Galle, who was well equipped to take on the job. Sure enough, Galle found a planet in the expected place, to within a degree or so of Le Verrier's and Adams's predicted values. Because Galle's observations came at Le Verrier's behest, most credit for the detection fell to him rather than to Adams.

This discovery was one of the most celebrated events in the history of science—the first time mathematics had been used to correctly ascertain the position of an unknown planet, opening up a whole new approach in astronomy. But matters did not end there, for it is not enough simply to pick out the right spot in the sky—or the right "ephemeris." One would also like to know the orbit of the body in question. And this is where Peirce entered the fray. The brash Yankee praised the work of Le Verrier and Adams, which led to Neptune's discovery, while suggesting that they got the right spot but the wrong planet, so to speak. Le Verrier initially believed that Neptune was about twice as massive as Uranus and lay about thirty-six astronomical units from the sun (one astronomical unit being the mean distance between Earth and the sun). After further analysis, Peirce supported the view that Neptune was less massive and less distant, lying thirty or so astronomical units from the sun—a conclusion drawn, in part, from the computations of Sears Cook Walker, an American astronomer based at the U.S. Naval Observatory. There was more than one possible solution, Peirce argued, including the one Le Verrier had originally advocated and the solution that Peirce had later come around to. In 1846, Peirce contended, these two solutions happened to line up in more or less exactly the same place, which is why he labeled Galle's discovery, based on Le Verrier's prediction, "a happy accident."[40]

This, as one might imagine, left Le Verrier none too pleased. He was among the world's preeminent mathematical astronomers, whereas Peirce, the wild-eyed American, was a relative unknown, especially in Europe, which ruled science in that day. Peirce took a bold stance that some would have characterized as outrageous. When Peirce announced at a meeting of the American Academy of Arts and Sciences in Cambridge that the discovery of Neptune had been accidental, Harvard president Edward Everett, who was present at the meeting, urged that a declaration so utterly improbable should not be presented to the world without the academy's backing. "It may be utterly improbable," Peirce replied, "but one thing is more improbable—that the law of gravitation and the truth of mathematical formulas should fail."[41]

Peirce held his ground without backing off. But the question remains: Was he right? Was Le Verrier really the beneficiary of a happy accident, as Peirce insisted? There are many ways of looking at this skirmish. In the end, Peirce's suggestion that Neptune was located about thirty astronomical units from the sun rather than thirty-six astronomical units turned out to be much closer to the truth. But Peirce's statements came after Le Verrier's and Adams's predictions and Galle's detection, as well as coming after subsequent work by Walker and others. Given the data initially available to Le Verrier and Adams, it simply was not possible to work out the orbital elements right off the bat. In the beginning, there is always a range of solutions—a range of possible distances and masses. One settles in on the precise orbit through an iterative process, after acquiring some data that show the planet's position at various junctures in history. Neptune's orbit, moreover, depended on Neptune's mass, which could not be calculated directly until a satellite of Neptune was discovered. In the end, many of Peirce's hunches were borne out, but that did not really take anything away from the achievements of Le Verrier and Adams. They worked out the ephemeris in an acceptable manner and could not possibly have pinned down the orbit from the outset. To some extent, both sides won, and none really lost.

Yet most Americans believed that Peirce came out ahead in the exchange, even if he had, in reality, just played to a draw.[42] Perhaps some validation can be drawn from the fact that four years after Galle's discovery, Peirce was admitted to the Royal Astronomical Society of London, the first American so elected since his mentor, Nathaniel Bowditch, had been similarly honored in 1818. That Peirce had taken on the scientific elite from Europe, emerging unscathed from those debates, "gave standing to both the scholar and his country," writes Emory mathematician and historian Steve Batterson. "The latter was important to him."[43]

As Hogan puts it, "Peirce was a scientific patriot. He saw the glory of America not in terms of Manifest Destiny or military might, but in terms of the nation's becoming a world leader in science and education." In the 1840s, many Americans were trying to win respect for their nation's scientists, and none worked harder to that end than Peirce himself.[44] Although Peirce was by no means lacking in ego, much of the work he did was not for self-aggrandizement, because he often did not bother to publish completed papers or, instead, let others take the credit. Beyond his individual accomplishments, Peirce was intent on showing that, when it came to science, Americans deserved a place on the world stage. Neither he nor his fellow countrymen were yet shaping mathematics on an international basis, but he was at least helping them get into the game.

With his credentials more firmly established in mathematical astronomy, in 1849 Peirce was named consulting geometer and astronomer to the *American Ephemeris and Nautical Almanac*. He made many and varied contributions to the publication over the next thirty years. For example, Peirce devised novel ways to take advantage of occultations of the Pleiades—instances when the moon obscures our views of the famous star cluster (also known as the Seven Sisters). He showed how detailed observations of the cluster, under these conditions, could reveal features about the shape and surface of both Earth and the moon.

In the 1850s, Peirce turned his attention to the rings of Saturn. Late in the eighteenth century, Laplace had suggested that Saturn had a large number of solid rings. In 1850, the American astronomer George P. Bond discovered a gap in the rings of Saturn during observations made with Harvard's great refractor. Bond believed the rings must be fluid rather than solid, as Laplace and others had maintained. Peirce undertook a detailed mathematical analysis of the rings' constitution, concluding that they were fluid. He showed, moreover, that the presence of Saturn alone would not keep the rings stable. But Saturn, along with its eight satellites, could keep a fluid ring in equilibrium. Peirce presented his findings in 1851 at a national science meeting in Cincinnati. A Boston newspaper lauded Peirce for putting forth "the most important communication yet presented" at the conference, as well as making "the most important contribution to astronomical science . . . since the discovery of Neptune."[45]

Unfortunately, Peirce's conclusion turned out to be incorrect. In 1859, the great physicist James Clerk Maxwell published a paper, "On the Stability of the Motion of Saturn's Rings," in which he argued that the rings were neither solid nor fluid but were instead composed of a countless number of small particles independently orbiting the planet. Maxwell's theory was confirmed in 1895, when astronomers James E. Keeler and William W. Campbell showed that the inner portion of the rings orbits more rapidly than the outer portion. Even though Peirce's idea did not prevail in the end, his analyses helped advance science by spurring Maxwell. In an 1857 letter to the physicist William Thomson (better known as Lord Kelvin), Maxwell wrote: "As for the rigid ring I ought to first speak of Prof. Peirce. He communicated a large mathematical paper to the American Academy on the Constitution of Ring, but up to the present year he has no intention of publishing it."[46] Lord Kelvin evidently had a high opinion of the aforementioned professor, for in an address before the British Association for the Advancement of Science (of which Lord Kelvin became president in 1871) he called Peirce the "Founder of High Mathematics in America."[47]

At this stage in his career, Peirce was already shifting his attention to geodesy—the branch of science concerned with measuring, monitoring, and representing Earth's shape and size, while also determining the precise location of points on the surface. From 1852 to 1867, Peirce served as director of longitude determinations for the U.S. Coast Survey—a job that technically involved making east-west measurements but was, of course, more broadly defined. At that time, Alexander Dallas Bache, a charismatic individual who dominated U.S. science in that era, headed the survey. Peirce and Bache became close friends, and when Bache died in 1867, Peirce took over his position as superintendent. A survey of Alaska, which had recently been purchased from Russia, was undertaken during Peirce's term—a period that saw a general improvement in scientific methods. Peirce surprised many people by his success at the survey's helm, given his lack of administrative experience, and he used his scientific reputation to secure more funding from Congress for basic research than even Bache had been able to obtain. Both Peirce and

Bache had established the U.S. Coast Survey as the most important federal agency for American science by supporting the research of people both inside the survey and outside of it.

Charles Sanders Peirce, Benjamin's son, worked intermittently on the survey from 1859 to 1891, but his accomplishments extended far beyond that. A brilliant polymath, Charles made contributions in mathematics, astronomy, chemistry, and other areas, but his principal achievements lay in the realm of logic and philosophy. Indeed, the philosopher Paul Weiss called Charles "the most original and versatile of American philosophers and America's greatest logician."[48] In some circles, Benjamin Peirce's greatest contribution to scholarly thought was bringing his son Charles into this world and helping to train him. While there is no denying that Charles made deep and lasting marks in his field—arguably leaving behind a greater and more durable legacy than that of his father—it is also true that Benjamin Peirce was himself one of those larger-than-life figures who could not help but attract the spotlight. And during his lifetime he cast a very big shadow indeed.

As an intellectual, with no false modesty about his cognitive abilities, the elder Peirce was not shy about sharing his opinion on all manner of subjects—science, arts, politics, and literature. As such, he did not feel obliged to restrict his pronouncements to mathematics alone. In one departure from his conventional duties with the math department and U.S. Coast Survey, Peirce attended a séance in Boston in 1857 to judge whether the participants could successfully communicate with spirits—a proposition of which he was highly skeptical. Peirce was there as an observer to appraise the validity of the proceedings. He was not surprised that the three-day event yielded no positive results. On a separate occasion, Peirce investigated the spiritualistic claims of a woman who said she came in contact with a universal force called "Od" in the presence of powerful magnets. In an experiment, Peirce showed those claims to be fraudulent: the woman exhibited the same reaction when exposed to a genuine magnet as she did when exposed to a piece of wood that was painted like a magnet.[49]

The séance- and spiritualism-busting activities were part of a broader effort that Peirce was involved in, along with other prominent friends and scientists, including Bache, as well as Louis Agassiz, an eminent zoologist and geologist at Harvard, and Joseph Henry, one of the country's leading scientists who served as the first secretary of the Smithsonian Institution. This group, which was part social club and part lobbying arm, called itself the Lazzaroni. Its principal aims were to rid American science of quacks and charlatans and, ultimately, to make the country the world leader in science. The name was intended to be humorous—a play on the Italian term *lazzaroni,* which referred to street beggars, since their American counterparts saw themselves as constantly begging to secure financial support for the nation's fledgling scientific establishment. The group's collective efforts led to the founding of the American Association for the Advancement of Science in 1848, of which Henry (1849), Bache (1850), Agassiz (1851), and Peirce (1852) all served as early presidents. The Lazzaroni also helped start the National Academy of Sciences, and Peirce was one of its most active early members.

In another departure from his usual routine, Peirce submitted testimony in 1867 as an expert witness in a celebrated court case. At issue was a will that left $2 million to Hetty Robinson, the niece of Sylvia Ann Howland, who died in 1865. The executor of the estate, Thomas Mandell, contested Robinson's claim, maintaining that the will was a forgery. Two of three signatures on the will, Mandell argued, had been traced. Peirce and his son Charles (who was then working on the U.S. Coast Survey with his father) both testified on Mandell's behalf, using statistical reasoning to demonstrate that the signatures were so close—with the "down strokes" matching so precisely—that the odds of their being genuine, rather than tracings, were one in 2,666,000,000,000,000,000,000. "Professor Peirce's demeanor and reputation as a mathematician must have been sufficiently intimidating to deter any serious mathematical rebuttal," wrote Paul Meier and Sandy Zabell in the *Journal of the American Statistical Association.* "He was made to confess a lack of any general expertise in judging handwriting, but he was not cross-examined at all on the numerical and mathematical parts of his testimony."[50]

Agassiz and Oliver Wendell Holmes testified for Robinson, saying they could find no signs of pencil marks that would have been evidence of tracing. Mandell prevailed in the end, although the extent to which the Peirces' arguments influenced the final verdict is not clear. (If nothing else, they are likely to have shaken confidence in the validity of the signatures on the will.) "Although Peirce's methods would be criticized by modern mathematicians, they were an early and ingenious use of statistical methods applied to a practical problem," Hogan writes. "Peirce's testimony may well be the earliest instance of probabilistic and statistical evidence in American law."[51]

While Peirce squared off against Agassiz on this particular occasion, they were united in their general desire to promote the national science agenda. Hogan believes that Peirce's scientific accomplishments, primarily in astronomy and mathematics, were overshadowed by "his efforts to organize American scientists into a professional body and to make educational reforms at Harvard . . . The development of an institutional base, centered in the universities, and the emergence of a professional scientific community were the most important developments in American science during the 19th century."[52]

Although Peirce devoted countless hours to this cause, he did not neglect his personal interests altogether. Of those, he considered pure mathematics his first love, even though he was not able to spend as much time on it as he might have liked. Indeed, only a small fraction of his published papers were in that area, with the bulk lying in more applied realms. That, however, may merely reflect the practical demands placed on his career, rather than his true intellectual leanings.

It is often said that a mathematician makes his or her most significant contribution early in life—typically by the age of thirty or so. Peirce defied the conventional wisdom in this area, as he did in many other areas, saving his greatest triumph in mathematics until 1870, when he hit the ripe age of sixty-one. That was the year in which he presented his treatise *Linear Associative Algebra.*

Apparently, Peirce had saved the best for last. That is the prevailing judgment of history, and Peirce felt that way as well. In a letter he sent in 1870, along with the manuscript, to George Bancroft, the U.S. ambassador to Germany and cofounder of Round Hill School (Peirce's first employer), Peirce spoke modestly of the enclosed work: "Humble though it be, it is that upon which my future reputation must chiefly rest."[53] In his dedication to *Linear Associative Algebra,* Peirce characterized the tract as "the pleasantest mathematical effort of my life. In no other have I seemed to myself to have received so full a reward for my mental labor in the novelty and breadth of the results. I presume that to the uninitiated the formulae will appear cold and cheerless. But let it be remembered that, like other mathematical formulae, they find their origin in the divine source of all geometry. Whether I shall have the satisfaction of taking part in their exposition, or whether that will remain for some more profound expositor, will be seen in the future."[54]

In some ways, *Linear Associative Algebra* seems to have come out of the blue, because Peirce had not done much original work in algebra before. In another sense, though, Peirce's efforts were not entirely surprising, since they grew out of Sir William Rowan Hamilton's invention of "quaternions" in 1843. Hamilton delivered his first lectures on quaternions in 1848, and Peirce was deeply impressed. "I wish I was young again," he said, though he was still in his thirties, "that I might get such power in using it as only a young man can get."[55] The subject clearly occupied his thoughts for many years, and he worked on it when he could, even interspersing his original contributions to algebra with his administrative responsibilities as head of the U.S. Coast Survey. "Every now and then I cover a sheet of paper with diagrams, or formulae, or figures," he told U.S. Treasury secretary Hugh McCulloch, "and I am happy to say that this work again relieves me from the petty annoyances which are sometimes caused by my receiving friends who do but upset."[56]

Before discussing Hamilton's work on quaternions and the effect it had on Peirce, it is worth saying a few words about algebra in general and how it began to change, and open up, in the early nineteenth century. That is when British mathematicians began transforming mathematics from the "science of quantity" to a much more liberalized and abstract system of thought. In the 1830s, for example, George Peacock of Cambridge University proposed that, in addition to arithmetical algebra, which involves the basic arithmetic operations and nonnegative numbers, there was also symbolic algebra: "The science which treats of the combinations of arbitrary signs and symbols by means of defined though arbitrary laws." Arithmetical algebra, Peacock averred, was really just a special case of the more general symbolic algebra.[57]

Hamilton carried this a step further by introducing complex numbers into his algebra. Complex numbers assume the form of $a + bi$, where a and b are real numbers and i, the square root of -1, is an imaginary number. Quaternions are four-dimensional representations of the form (a,b,c,d) or $a + bi + cj + dk$, where a, b, c, and d are real and i, j, and k are imaginary. These numbers obey various rules, such as $i^2 = j^2 = k^2 = -1$, and $ij = -ji$. Whereas Peacock held that the same rules applied to both symbolic and arithmetic algebra, this was not the case in Hamilton's system: in arithmetic algebra, $a \times b$ is always equal to $b \times a$, in adherence to the commutative law of multiplication, but the commutative law does not always apply to quaternions, since $i \times j$, by definition, does not equal $j \times i$. Hamilton believed that algebraists were not bound to set rules but were instead free to write their own rules as they saw fit. "Hamilton's work on quaternions revealed what has since come to be known as the freedom of mathematics, essentially the right of mathematicians to determine somewhat arbitrarily the rules of mathematics," writes Helena Pycior, a historian at the University of Wisconsin, Milwaukee.[58]

Fascinated by quaternions, Peirce discussed them in a course in 1848 (the year of Hamilton's first lectures on the topic), and he often called them his favorite subject. If anything, he was too enamored with quaternions, in the opinion of his son Charles, who complained that his father was a "creature of feeling" with "a superstitious reverence for the square root of minus one."[59] But Benjamin's preoccupation eventually paid off. He identified and provided multiplication tables for 163 different algebras up to the "sixth order," that is, containing six or fewer terms. These algebraic systems obeyed the associative law, $a(bc) = (ab)c$, and the distributive law, $a(b + c) = ab + ac$, but not the commutative law. Of these 163 algebras, only three had been in common use at the time: ordinary (arithmetic) algebra, the calculus of Newton and Leibniz, and Hamilton's quaternions.

Peirce went beyond Hamilton by insisting that the coefficients of his algebras, such as a, b, c, and d, could be complex numbers and need not be limited to real numbers. "Peirce became so infected with the freedom of mathematics," Pycior writes, that he accused Hamilton of "mathematical conservatism." As a result, his algebras "diverged even farther from arithmetic than the quaternions."[60]

Peirce had a great insight, demonstrating that "in every linear associative algebra, there is at least one idempotent or nilpotent expression."[61] A nilpotent is defined as the element a for which a positive integer n (greater than or equal to 2) exists such that $a^n = 0$. An idempotent is the element b for which a positive integer m (greater than or equal to 2) exists such that $b^m = b$.

The nilpotent was a somewhat controversial notion in algebra since a, by definition, is a divisor of zero, which is forbidden in the standard arithmetic version of algebra. Divisors of zero are nonzero numbers, a and b, such that $a \times b = 0$. Hamilton's system did not admit the possibility of divisors of zero, whereas Peirce's algebras allowed for it and were therefore even more general. (The introduction of divisors of zeros was, in fact, a consequence of Peirce's use of complex coefficients rather than just real number coefficients.) As if anticipating criticism on this score, Peirce wrote in his paper:

> However incapable of interpretation the nilfactorial and nilpotent expressions may appear, they are obviously an essential element of the calculus of linear algebras. Unwillingness to accept them has retarded the progress of discovery and the investigation of quantitative algebras. But the idempotent basis seems to be equally essential to actual interpretation. The purely nilpotent algebra may therefore be regarded as an ideal abstraction, which requires the introduction of an idempotent basis to give it any position in the real universe.[62]

By introducing new concepts in such a comprehensive manner, Peirce laid out a broad new terrain for future study—scores of algebras that had never before been considered, let alone explored. Through that accomplishment, Peirce laid "a just claim to be considered an eminent mathematician," according to George David Birkhoff, whom many regarded as Harvard's dominant mathematician in the first half of the twentieth century, as well as, by many accounts, the greatest American mathematician of his time.[63] "Peirce saw more deeply into the essence of quaternions than his contemporaries, and so was able to take a higher, more abstract point of view, which was algebraic rather than geometric." Peirce thus appears, Birkhoff adds, "as a kind of father of pure mathematics in our country."[64] That said, it should be stressed that the most groundbreaking work in the field at the time was still, overwhelmingly, occurring in Europe.

As a historian, Pycior agrees with Birkhoff's appraisal, calling *Linear Associative Algebra* "a pioneer work" in American mathematics. "Peirce deserves recognition not only as a founding father of American mathematics," she writes, "but also as a founding father of modern abstract algebra."[65]

The recognition that Peirce received for his work in algebra, like that bestowed by Birkhoff and Pycior, was long in coming, partly because of the manner in which his results were communicated to the mathematical community. At first Peirce presented his work orally, reading his "memoir" before the National Academy of Sciences in 1870, with earlier installments coming before then. This was not the ideal way of conveying such abstruse material. Agassiz, who sat through an earlier presentation on the subject, spoke for other confused audience members in saying: "I have listened to my friend with great attention and have failed to comprehend a single word of what he has said. If I did not know him to be a man of great mind . . . , I could have imagined that I was listening to the vagaries of a madman."[66]

Peirce had a better chance of getting his message across by publishing his paper, but that did not happen—except on an extremely limited basis—during his lifetime. The National Academy of Sciences intended to publish it but never got around to it. One hundred lithograph copies were produced, however, with the help of the U.S. Coast Survey staff. The work was done, in particular, by "a lady without mathematical training but possessing a fine hand . . . who could both read his ghastly script and write out the entire text 12 pages at a time on lithograph stones."[67] Most of the copies were sent to Peirce's U.S. colleagues and friends, who, unfortunately, lacked the expertise to appreciate his accomplishment. The paper had a reasonably good reception in England, where William Spottiswoode, the outgoing president of the London Mathematical Society, summarized Peirce's results in an 1872 talk to the society. But Peirce was unable to get the mathematical community in Germany—then the world leader in the field—to take stock of his work.

In 1881, a year after Peirce died, *Linear Associative Algebra* finally appeared in its entirety in a more accessible venue, the *American Journal of Mathematics*—an outcome that occurred through the initiative of his son Charles. An introduction described the work as one that "may almost be entitled to take rank as the *Principia* of the philosophical study of the laws of algebraic operation," thus comparing Peirce's feat—perhaps hyperbolically—to Isaac Newton's famous treatise, which contained his laws of motion and gravitation.[68]

Linear Associative Algebra was Peirce's first major contribution to pure mathematics, as most of his work prior to that time had been in astronomy, physics, and geodesy. It was, arguably, the first major contribution to mathematics by an American at all. Bowditch's translation of Laplace, though a prodigious feat, was still, at heart, an explanation of another's work. And Peirce's paper on odd perfect numbers was not of the same stature, since it merely put a constraint—albeit the first constraint—on the problem, unlike his later paper, which helped lay the groundwork for future inquiries into abstract algebra.

Ironically, just as Sylvester and Servais had duplicated Peirce's efforts on odd perfect numbers, because his work was not widely known, so too were much of Peirce's algebraic efforts duplicated twenty years later by two German mathematicians, Eduard Study and

Georg Scheffers, who had either overlooked Peirce's paper or not taken it seriously. In two papers published in 1902, Columbia University mathematician Herbert Hawkes maintained that the theorems stated by Peirce "are in every case true, though in some cases his proofs are invalid." Hawkes went through the proofs, making corrections or clarifications when necessary, to place the entire work "on a clear and rigorous basis ... Using Peirce's principles as a foundation," he argued, "we can deduce a method more powerful than those hitherto given for enumerating all number systems of the types Scheffers has considered." The reason Peirce's memoir was "subject to neglect or adverse criticism," Hawkes speculated, owed in part to "the extreme generality of the point of view from which his memoir sprang, namely a 'philosophic study of the laws of algebraic operation.'"[69]

Writing in 1881, the Yale mathematician Hubert Anson Newton also suggested that while the paper broke new ground, it definitely had a philosophical edge, offering what he thought would become "the solid basis of a wide extension of the laws of thinking."[70]

It is doubtful that Peirce would have resented the characterization of his work as "philosophic," since the very first sentence of the paper—as well as the first two pages—provides a general discussion of what mathematics is all about. He starts the discourse by writing,

> Mathematics is the science which draws necessary conclusions. This definition of mathematics is wider than that which is ordinarily given, and by which its range is limited to quantitative research ... The sphere of mathematics is here extended, in accordance with the derivation of its name, to all demonstrative research, so as to include all knowledge strictly capable of dogmatic teaching ... Mathematics, under this definition, belongs to every enquiry, moral as well as physical. Even the rules of logic, by which it is rigidly bound, could not be deduced without its aid.[71]

Peirce thus rejected the notion that mathematics is merely the science of *quantity* in favor of the much broader notion of mathematics being a science based on inference and deduction. In an earlier address to the American Association for the Advancement of Science, Peirce called mathematics "the great master-key, which unlocks every door of knowledge, and without which no discovery—no discovery, which deserves the name, which is law and not isolated fact—has been or ever can be made."[72]

Peirce's views about mathematics were deeply colored by his fervent religious convictions. He considered mathematics one of the highest forms of human expression and, as such, a manifestation of God's infinite wisdom. Like Plato and Aristotle before him, Peirce believed that "God wrote the universe in the language of mathematics."[73] Peirce made no attempt, moreover, to conceal his religious feelings and was, instead, quite open about them. Indeed, in the introductory paragraph to his 1870 paper, he noted that the mathematical formulas contained therein had a divine origin—something that he considered to be true of mathematics in general. While hoping to have a role in the "exposition" and advancement of these ideas, he admitted that his ultimate contribution to this effort remained to be seen.[74]

As it turned out, Peirce did not participate much in the further investigation of the algebras that he had laid out, so systematically, in his paper, though the "exposition" of which he spoke has been central to the development of modern abstract algebra. Peirce clearly had other interests in mind, though some were related to his work on algebraic systems. He gave a series of lectures late in his life, posthumously published in a volume entitled *Ideality in the Physical Sciences,* in which he argued that every physical phenomenon could be expressed through mathematics and, conversely, that every mathematical idea had an expression in the physical world.[75] "There is no physical manifestation which has not its ideal representation in the mind of math," he affirmed.[76] During the 1879–80 academic year, Peirce was engaged in the study of cosmology—or "cosmical physics," as he called it. He planned to teach a course on the subject in the following year, but his health failed him. Peirce died on October 6, 1880, before he had the chance to explain the universe—its origins, formation, and evolution—to those Harvard students capable of following his often vexing style of discourse.

Peirce approached death stoically, girded by his lifelong faith. He expressed no great sorrow, for instance, when his father died in 1831 at the age of fifty-three. Had he lived longer, Peirce wrote to his father's physician, "would he have been happier? Thank God, no. He is in heaven, and I will not regret that no human power could ward off that fatal blow."[77]

Peirce was similarly resigned to the prospect of his own mortality. "Distinguished throughout his life by his freedom from the usual abhorrence of death, which he never permitted himself either to mourn when it came to others, or to dread for himself, he kept this characteristic temper to the end," wrote F. P. Matz, a professor of mathematics and astronomy at New Windsor College in Maryland, in 1895. "Two days before he ceased to breathe, it struggled into utterance in a few faintly-whispered words, which expressed and earnestly inculcated a cheerful and complete acceptance of the will of God."[78]

Although Peirce may not have feared the exercise of God's will, when it finally came to him, he was apprehensive about the accolades that were likely to be bestowed upon him after his death. More than a decade and a half earlier, in the spring of 1864, Peirce thought he was terminally ill, requesting of his friend Bache that, "if I should be taken, dearest Chief, exert all your influence to save me from eulogistic biographers."[79] But Peirce's worst fears were realized on this score when the end came in 1880, with eulogies and memorial tributes pouring in from all quarters. In fact, an entire book (albeit a small one), *Benjamin Peirce: A Memorial Collection,* was published a year later to honor the man who had served Harvard for nearly half a century. The contents of the volume, the editor wrote, "but feebly reflect the life of one who ranks among the few men whose names have been imperishably recorded in the annals of science and religion in this century." The collection contained poems, sermons, tributes, and obituaries from various newspapers and magazines, including the *Springfield Republican,* which declared: "America has nothing to regret in his career but that it must now be closed; while her people have much to learn from his long and honorable life."[80]

In 1880, the *Atlantic Monthly* published a poem by Peirce's Harvard classmate Oliver Wendell Holmes:

> Through voids unknown to worlds unseen
> His clear vision rose unseen . . .
>
> How vast the workroom where he brought
> The viewless implements of thought!
>
> The wit, how subtle, how profound,
> That Nature's tangled webs unwound.[81]

"By the death of Professor Benjamin Peirce, last week, the University loses its greatest light in science, and perhaps the most distinguished of its professors," the *Harvard Crimson* wrote.[82] In the wake of Peirce's death, the Harvard mathematics department entered a "period of retrocession," according to Coolidge, "a great slump in . . . scientific activity" that would take years to dig out of. The good news, he wrote, is that "a renaissance" of mathematics at Harvard would come more than a decade later, led by newly appointed faculty members whom he referred to as the "great twin brethren . . . A momentous revival in American mathematics," as Coolidge put it, was about to begin.[83]

2

OSGOOD, BÔCHER, AND THE GREAT AWAKENING IN AMERICAN MATHEMATICS

As the first person to conduct significant mathematical research at the school, Benjamin Peirce was truly a Harvard pioneer, even if he was practically the only one doing such work, and even if he had to fit in those efforts between the chores that actually paid his bills: teaching and writing textbooks. The next leap forward for the Harvard mathematics department was made around the turn of the twentieth century by William Fogg Osgood and Maxime Bôcher, who turned Harvard into a powerhouse in the field of analysis—a branch of pure mathematics that includes calculus, as well as the study of functions and limits. But from a long-term perspective, two other accomplishments of Osgood and Bôcher may loom even larger: they brought mathematical research to the center of the department's mission rather than just being the hobby of an incorrigible nonconformist like Peirce, and what is more, they transformed Harvard's department into what was arguably the strongest in the country, in the face of considerable competition, during a time of great progress in American mathematics.[1]

That transition did not come easily, however. Following Peirce's death in 1880, Harvard mathematics underwent a major decline on the research front (although perhaps not on the teaching front, since molding young minds—especially rather ordinary young minds—was never one of Peirce's fortes). "The state of mathematical scholarship at Harvard in the 1880s . . . had reverted back to that at the beginning of the century," explains Steve Batterson, a mathematician and math historian at Emory University. "No one was proving new theorems."[2]

These repercussions extended beyond the department itself, given that, for the better part of the century, Peirce had been the country's leading mathematical researcher. And even then, his most important mathematical work was not widely known until after his death. Nor did Peirce and his colleagues produce a successor generation of mathematical researchers, in part because the infrastructure to train such people did not really exist at the time. In 1875, five years before his father's death, Charles Sanders Peirce held that Harvard did not "believe in the possibility of any great advances in science . . . being made there." The prevailing assumption, he added, was that "the highest thing it can be is a school."[3]

Speaking more broadly in 1874, the mathematical astronomer Simon Newcomb, a former student of Peirce's at Harvard, noted that "the prospect of mathematics [in the United States] is about as discouraging as the retrospect." An honest appraisal of the quantity and quality of original published research should prompt us, Newcomb said, "to contemplate the past with humility and the future with despair." The problem, as he saw it, owed to the "lack of any sufficient incentive to the activity which characterizes the scientific men of other nations, and of any sufficient inducement to make young men of the highest talents engage in science research."[4] Newcomb's grim assessment was made a couple of years before the founding of the *American Journal of Mathematics,* which meant that at the time of his statement, a publication dedicated to mathematical research did not exist—nor had ever existed—anywhere in the country. "Even if there had been places to publish," writes historian Karen Hunger Parshall, "there was no shared sense that the advance of knowledge and the communication of that knowledge through publication represented desired ends."[5]

Charles Eliot, a former Harvard professor of mathematics and chemistry who became president of the university in 1869, was well aware of this problem. Following the deaths of two prominent faculty members—the zoologist Louis Agassiz and the anatomist Jeffries Wyman, in 1873 and 1874, respectively—Eliot wrote that "the want of Americans worthy to take their places, or those of Asa Gray and Benjamin Peirce, was a significant proof of the failure of American universities to breed scholars, and the strongest possible argument for a graduate school," as paraphrased by Samuel Eliot Morison.[6] Eliot created a graduate department in mathematics that was authorized to grant M.A. and Ph.D. degrees. The first Ph.D., as mentioned in Chapter 1, was awarded in 1873 to the mathematician William Elwood Byerly, the only student to earn a Ph.D. under Peirce's tutelage. The graduate program that existed at the time was not geared to turning out students ready for original research, despite Eliot's goal of transforming Harvard into a modern university. Although Eliot made progress in many Harvard departments, change was slow to come in the mathematics department.

An important milestone occurred in 1876, the nation's hundredth anniversary, with the founding of Johns Hopkins University—the nation's first research-oriented university. A year later, the distinguished British mathematician James J. Sylvester—a pioneer in invariant theory, among other claims to fame—was put at the head of the school's mathematics department, with the principal emphasis placed on graduate work, though not to the detriment of undergraduate studies. American students finally had opportunities to pursue mathematics in ways that were not available to them before. Although Johns Hopkins was the first university of its kind in the United States, Peirce and his Lazzaroni friends had dreamed of creating just such an institution around 1850 or so. In their fantasies, however, the new university was to reside in Albany, New York, rather than in Baltimore, Maryland, but those plans never materialized. The establishment of Johns Hopkins, a quarter of a century later, was widely seen as an important step in the advancement of American mathematics, though it produced a temporary setback for Harvard's fledgling graduate program as the new university started attracting many of the best graduate students.

Peirce contributed, indirectly, to Harvard's graduate student shortfall by writing a letter in support of Sylvester to Johns Hopkins's first president, Daniel Gilman:

> I take the liberty to write you concerning an appointment in your new university, which I think would be greatly for the benefit of our country and of American science if you could make it. It is that of one of the two greatest geometers of England, J. J. Sylvester. If you enquire about him, you will hear his genius universally recognized but his power of teaching will probably be said to be quite deficient. Now there is no man living who is more luminary in his language, to those who have the capacity to comprehend him, than Sylvester, provided the hearer is in a lucid interval. But as the barn yard fowl cannot understand the flight of the eagle, so it is the eaglet only who will be nourished by his instruction ... Among your pupils, sooner or later there must be one who has a genius for geometry. He will be Sylvester's special pupil—the one pupil who will derive from his master, knowledge and enthusiasm—and that one pupil will give more reputation to your institution than the ten thousand, who will complain

of the obscurity of Sylvester, and for whom you will provide another class of teachers ... I hope you will find it in your heart to do for Sylvester what his own country has failed to do—place him where he belongs—and the time will come when all the world will applaud the wisdom of your selection.[7]

Apart from respecting Sylvester for his mathematical prowess, Peirce may also have been sympathetic because his own teaching style, like Sylvester's, was regarded as idiosyncratic at best—an acquired taste (to be charitable) that did not go over well with the masses.

After some negotiations over salary, Sylvester accepted the job. The next issue to be taken up was that of faculty appointments—the question being who was capable of providing the proper leadership at a research-driven school. When asked for his opinion, Peirce submitted that he and Sylvester were perhaps the only people in the country qualified to select suitable candidates—the exact problem he and others had hoped to remedy through the creation of a school like Johns Hopkins. Based on Peirce's recommendation, the Harvard tutor William Story, who got his undergraduate degree at Harvard and his Ph.D. in Leipzig, Germany, was picked to teach geometry and serve as Sylvester's second in command.

Part of the Johns Hopkins philosophy, in keeping with the European model, was that the research of faculty and students should be geared, whenever possible, toward publication in respected journals. To further this end, and to help promote mathematics in the country as a whole, Sylvester, with help from Story, Newcomb, Peirce, and others, started the aforementioned *American Journal of Mathematics* in 1878. The journal's avowed purpose, unlike that of previous or existing American math-related periodicals, was for "the publication of original investigations."[8] That is where Peirce's *Linear Associative Algebra* was eventually published, and had an outlet like that existed before, his magnum opus might have attained some readership while he was still alive.

Sylvester left Johns Hopkins in 1883 to secure a chaired faculty spot at Oxford. Simon Newcomb took over his position at Johns Hopkins. But since Newcomb was also serving as superintendent of the *Nautical Almanac* office in Washington, D.C., he could spend only two days a week in Baltimore, where he mainly taught astronomy. "It was not enough to make up for the loss of Sylvester," writes Batterson, noting that the department's subsequent decline was almost as swift as its rise to prominence. "Once again, there was no United States university providing mathematical training approaching what was available in Europe."⁹

In the wake of Sylvester's departure, the Johns Hopkins mathematics department lost its unofficial title as the nation's leading program, but the venture had been successful nevertheless. "The establishment at Hopkins of a graduate school which engaged in properly graduate education, that is, in the training of future researchers, forced other institutions which saw more advanced education as part of their mission, to establish similar schools," wrote Parshall and David E. Rowe. "The guiding philosophy of Johns Hopkins, with its emphasis on graduate training and research, was transferred to extant universities, like Harvard, Princeton, and Yale and to newly forming ones" as well.¹⁰

Of the new universities referred to above, the most important at the time, insofar as mathematics was concerned, was the University of Chicago, whose mathematics department was placed under the direction of E. H. (Eliakim Hastings) Moore, an American who got his Ph.D. from Yale in 1885 before spending an additional year of study in Germany. Although Moore made important contributions to geometry and group theory, he is remembered more for his success at building Chicago's department, which had a galvanizing effect on mathematics nationwide. When the school opened its doors for the first time in the fall of 1892, Moore had two able mathematicians at his side, Oskar Bolza and Heinrich Maschke, both of whom were from Germany and students of Felix Klein's. Chicago got off to a fast start, turning out large numbers of math Ph.D. recipients. Some of the most prominent ones trained by Moore include George David Birkhoff, soon to become a forceful presence at Harvard; Leonard Eugene Dickson, the program's first Ph.D. recipient, who spent most of his career on the Chicago faculty; and Oswald Veblen, a gifted geometer who settled at Princeton, where he became a leading American mathematician. Many believed that Moore was "the prime driving force which finally turned the United States from a mathematical wasteland into a leader in the field," Parshall wrote.¹¹

Few would deny Moore's impact, but while happenings at the brand new Chicago campus were grabbing attention, quieter—though equally significant—changes were under way at other American universities, including its first and oldest. Simply put, the examples set by Johns Hopkins and Chicago—by stressing that its professors conduct research and prepare students to do the same—led to a pronounced upswing in American mathematics by the turn of the twentieth century. The best institutions came to realize not only that they could change their ways but also that they had to. These schools, bound by long-held traditions, were at a disadvantage compared with Chicago, which was set up to foster research and graduate studies from the outset. The established schools, in other words, had to work harder to keep pace.

Harvard was among the places that managed to do so, overcoming centuries of inactivity in mathematical research. The school's mathematics department came of age in this period, bolstering its international reputation, thanks largely to the presence of the two young faculty members introduced earlier, Osgood and Bôcher, who started as instructors in 1890 and 1891, respectively, and moved steadily up the ranks to full professorship. Both Osgood and Bôcher were Boston-born graduates of Harvard College who returned to the university after earning doctoral degrees in Germany. And both had come back from their overseas experience with a well-earned confidence that they were the equals of their European counterparts and that, like them, other American students should be capable of the same. They left a permanent mark on the department, both as research mathematicians and as teachers, setting high scientific standards for themselves, their colleagues, and their students. They also left a permanent mark on the field of analysis, their mathematical specialty, establishing Harvard as a center for such research.

"It was the moment of the great awakening in American mathematics, when a number of able young men, largely trained in Germany, set about raising the science as pursued in this country to the same plane on which it was pursued in Europe," explained Harvard mathematician Julian Coolidge. Osgood and Bôcher, whom Coolidge called the "great twin brethren . . . were in the very middle of it," bringing new and advanced courses into Harvard and surrounding themselves with a core of talented, enthusiastic students.[12] Most important, these new instructors instilled a new attitude in their students and fellow faculty members— one that Benjamin Peirce had embraced as well: that practitioners in the field needed to contribute to the advancement of mathematical knowledge. They helped bring about the blossoming of mathematics nationwide by the example they set—helping to build a strong department at Harvard and also publishing first-rate research in journals based in the United States and abroad. In a 1903 ranking of the top eighty American mathematicians, Osgood and Bôcher made the top four, alongside Moore and George William Hill (the latter having worked for Benjamin Peirce at the *Nautical Almanac* office in Cambridge).

Like many who go on to noteworthy careers in mathematics, neither Osgood nor Bôcher knew exactly what they wanted to do with their lives upon first entering Harvard as undergraduates. Both, in fact, had exceedingly broad interests. Bôcher, the son of a Harvard French professor, spent his undergraduate years studying Roman and medieval art, music, chemistry, geology, geography, meteorology, philosophy, and zoology, before ultimately settling on mathematics. He won top honors in the latter field for his thesis on systems of parabolic coordinates. In such a system (in the two-dimensional case) the coordinate lines consist of two orthogonal parabolas that are "confocal," that is, sharing the same focus.

Osgood entered Harvard intent on studying the classics—Greek authors, in particular—and for the first two years of college he devoted himself to that aim. "There was little in the Harvard curriculum at the time to inspire a young man to give his life to mathematical research," wrote Coolidge, Birkhoff, and Harvard physicist Edwin C. Kemble in a tribute to Osgood. "The only member of the mathematics department actively interested in scientific advance was the youthful Benjamin Osgood (B. O.) Peirce, and his interest lay largely in the field of physics. But Osgood had early absorbed the idea that mathematics was the most difficult subject to be studied, and he meant to try for the biggest prize."[13]

B. O. Peirce—a distant relative of Benjamin Peirce and a Hollis chair holder from 1888 to 1914—was one of Osgood's favorite teachers. He, along with the algebraist Frank Nelson Cole, who started lecturing in the department during Osgood's senior year, helped spark Osgood's interest in math. Cole played an especially instrumental role. After graduating from Harvard in 1882, he went to Germany to study with Felix Klein, who pioneered the concept of Kleinian groups and established new and stronger ties between geometry and symmetry. Cole returned to Harvard, finishing up his Ph.D. work in 1886. He was able to afford the trip to Germany thanks to a $1,000-a-year Parker fellowship, named after the Boston merchant John Parker Jr., who established the fellowship to allow top Harvard students to continue their studies after graduating from college.

With his graduate work complete, Cole spent two years as a lecturer at Harvard before moving to the University of Michigan. While at

Harvard, Cole attracted the attention of two undergraduates, Osgood and Bôcher, both of whom were inspired by his accounts of studying in Germany under Klein, who was then the most popular mathematician for Americans seeking higher education abroad. As a measure of Klein's success, six of his students, including Osgood and Bôcher, later served as presidents of the American Mathematical Society. The society was founded in 1894 as an outgrowth of the New York Mathematical Society, charged with promoting mathematical research and scholarship in the country.

At the time, Klein was widely considered to be the most charismatic teacher of mathematics in the world. He had a gift for communication, combined with an extraordinarily broad scientific knowledge base, as well as a knack for identifying promising new avenues. Perhaps his greatest strength was in training students to conduct independent research of their own rather than following in his footsteps in geometry and other areas of mathematics.

As a lecturer at Harvard, Cole was eager to share the "new math" he had learned regarding Riemann surfaces (one-dimensional complex surfaces first investigated by the eminent German mathematician Bernhard Riemann), group theory, and other topics. He was "aglow with the enthusiasm which Felix Klein stirred in his students," Osgood reported, calling that enthusiasm contagious. As Osgood saw it, the regular faculty members—James Mills Peirce, William Byerly, and B. O. Peirce—"stood as the Old over against the New, and of the latter Cole was the apostle. The students felt that he had seen a great light. Nearly all the members of the Department . . . attended his lectures. It was the beginning of a new era in graduate instruction in mathematics at Harvard, and mathematics has been taught here in that spirit ever since."[14]

Osgood was obviously fired up by Cole's presentations, and Cole urged him to travel overseas to the University of Göttingen to study with Klein. Unable to get a Parker fellowship, Osgood obtained support from a so-called Harris fellowship instead. Bôcher followed suit a year later, also the beneficiary of a Harris fellowship. He, too, went to Göttingen, where he studied for six semesters under Klein.

Another student of Klein's that Osgood met during his first year in Germany was Harry Tyler, who was on a two-year leave from the MIT faculty. Tyler decided to spend his second year abroad in Erlangen, working under the guidance of two strong mathematicians, Paul Gordan and Max Noether (the latter being the father of Emmy Noether, a leading algebraist who is discussed in Chapters 4 and 7). Tyler advised Osgood to do the same after Osgood had spent two years with Klein. Although he had a high opinion of Klein, Tyler said that "so busy a man can not and will not give a student a very large share of his time and attention; so too he will not study out or interest himself in the painstaking elaboration of details, preferring to scatter all sorts of seeds continually and let other people follow after to do the hoeing." Noether, by contrast, Tyler said, "will probably have but one student besides myself and will probably give us anything we like . . . Now as to your plans, I would advise you unhesitatingly to come here if you want detailed work in pure mathematics."[15]

Osgood took Tyler's advice, spending his third year in Erlangen, where he earned his Ph.D. under the supervision of Noether. His dissertation was on so-called abelian integrals—integrals along certain curves on the complex plane, named after the Norwegian mathematician Niels Abel, that have been important tools in number theory and algebraic geometry. The main thrust of Osgood's thesis, which built on the previous work of Klein and Noether, fell within the realm of function theory—a subject that Osgood learned from Klein to which he would later dedicate much of his working life.

Osgood drew on function theory, for example, in his proof of the Riemann mapping theorem (discussed later in this chapter). Former Harvard math professor Joseph Walsh considered this paper Osgood's greatest achievement.[16] Osgood also wrote a 1907 textbook on function theory that remained a classic of the field for decades. "The years spent in Germany determined absolutely his whole future life," wrote Coolidge, Birkhoff, and Kemble. "In Germany he had such a large vision of the sort of work he would like to do that its accomplishment and natural extensions sufficed for the whole of his productive life."[17]

There might have been another reason for Osgood's moving to Erlangen. According to a story that Walsh heard, "Osgood became so enamored of a Göttingen lady that his work suffered, and Klein sent him to Erlangen for his doctorate." Regardless of which version of the story is correct, Walsh adds, Osgood got his degree from Erlangen in 1890, "and one or two days later he married the girl in Göttingen, and one or two days still later, they sailed for the United States of America."[18]

He took his new degree and new German wife back to Harvard, where he was installed in his new position as instructor of mathematics. Like many of his American peers, who had returned to their country after studying mathematics in Germany, Osgood was eager to boost the level of mathematics at home. He observed little mathematical research going on at Harvard at the time, apart from his own, but found a kindred spirit in Bôcher, who joined the faculty a year later.

The university they had returned to, after their German sojourn, was "rather like that of a provincial college," wrote Bernard Osgood Koopman, a mathematician (and relative of Osgood's) who earned his Ph.D. at Harvard under Birkhoff's supervision. "It had contained individuals of eminence, but it could scarcely have supplied a real training in advanced modern mathematics."[19] Together, Osgood and Bôcher supplied the critical mass needed to foster a new attitude in the department and to overhaul the prevailing culture. Perhaps the most obvious change was a fairly rapid upswing in research activity. Both young professors made it a high priority to carve out time to publish scientific papers of high quality, and they did so at a rapid clip: by 1900, Osgood had published twenty-one scientific papers, while Bôcher had published thirty papers, plus a survey article and one textbook, *On the Series Expansions of Potential Theory.*[20]

Osgood's first publication of note was an 1897 paper on the convergence of sequences of continuous functions of real variables. A "continuous" function, f, is one that makes no wild jumps—a function for which small changes in input result in small changes in output. Another way of putting it is that if you pick two points, x and y, that are close together, then $f(x)$ and $f(y)$ should be close together, too, assuming f is continuous. The next step is to consider a sequence of continuous functions, $f_1, f_2, f_3, \ldots, f_n$, which converge to a continuous function, f, as n tends toward infinity. One might then ask whether the areas under the curves $f_1, f_2, f_3, \ldots$ converge to the area under f. This is really a question about integration—a mathematical technique for describing, and quantifying, the area under a curve: does the integral of f_n over the interval from a to b converge to the integral of f over the same interval? Generally speaking, the answer to that question is no: the area under f_n does not converge to the area under f.

Osgood proved, however, that the area under f_n does converge to the area under f, provided that the sequence $f_1, f_2, f_3, \ldots$ is bounded. Saying that a sequence is bounded means that, for all values of n and x, the absolute value of $f_n(x)$ does not exceed M for a fixed and positive (though arbitrary) M: $|f_n(x)| \leq M$ for all n and x.

Osgood's result, when extended to noncontinuous or discontinuous functions, served as a model for the new method of integration put forth by the French mathematician Henri Lebesgue in 1907.[21] Osgood might have taken that step himself had he accepted "the striking consequences of his own conception," argued the mathematician and former child prodigy Norbert Wiener, who got his Ph.D. from Harvard in 1912 at the age of eighteen. (Wiener later taught philosophy courses at Harvard before accepting a long-term faculty post in mathematics at MIT.) Osgood, said Wiener, "must have had some rankling awareness of how he had missed the boat, for in his later years he would never allow any student of his to make use of the Lebesgue methods."[22]

Despite Osgood's possible disappointment over not having taken on the case of discontinuous functions, his 1897 effort was still "among the first papers by an American to receive international attention as part of the mainstream mathematical dialogue in Europe," according to Diann Renee Porter, a math scholar formerly at the University of Illinois at Chicago.[23]

Osgood's next big paper, which came in 1900, really put him on the map, so to speak. In this work he proved the Riemann mapping theorem—a high-profile problem that was first posed by Riemann in 1851. As the name implies, the theorem relates to the notion of a map—a function being one kind. A central concept in mathematics, a map, essentially, is a set of rules for taking a mathematical object, such as X, and assigning to each point in X a point in another mathematical object, Y.

In this particular case, Riemann proposed that a simply connected region of the plane, such as the region bounded by a simple closed curve, can be mapped one-to-one and conformally onto a disk, which is the region bounded by a circle. The mapping is "one-to-one" because every point bounded by the arbitrary curve corresponds to a single point on the disk and vice versa. The mapping is called "conformal" because angles are preserved. To illustrate what that means, suppose that two lines intersect at a particular angle within the domain bounded by the curve. When this domain is mapped conformally onto a disk, those two lines will correspond to two curves in the disk that intersect each other at the same angle.

Osgood was often drawn to problems of this nature, explained Walsh, "problems that were both intrinsically important and classical in origin—'problems with a pedigree,' as he used to say."[24] And his proof of Riemann's theorem stands as a major accomplishment. Some of Europe's greatest mathematicians, including Henri Poincaré, had attempted previously, though unsuccessfully, to find a proof, said Walsh. "This theorem remains as Osgood's single outstanding result."[25] Generalizing the Riemann mapping theorem to three dimensions, rather than two, essentially gives rise to the Poincaré conjecture, which stood for a century as one of the most intractable problems in mathematics, until the Russian mathematician Grigori Perelman—building on the work of Richard Hamilton from Columbia University—posted a series of papers on the web in 2002 and 2003.

The two problems, famously posed by Riemann and Poincaré, are linked because the Riemann mapping theorem, which Osgood proved, established the topological equivalence between the two-dimensional region of a plane bounded by a simply connected (nonintersecting) curve and a disk. The Poincaré conjecture, issued some fifty years later, proposed an equivalence between a simply connected three-dimensional space (that is "closed" or finite in extent) and a sphere. (One key difference is that Riemann's theorem concerns conformal mappings, which preserve angles. The Poincaré conjecture, on the other hand, concerns "homeomorphisms"—mappings that are continuous and one-to-one, with every point in the original space corresponding to a single point in the image space.)

Osgood came up with another interesting result in his 1903 paper "A Jordan Curve of Positive Area." A Jordan curve is a continuous closed curve, as described above, sitting in a plane that does not intersect itself. Named after the French mathematician Camille Jordan, a Jordan curve (based on a difficult theorem proved by Jordan and, separately, by Oswald Veblen) divides the plane into two regions: a compact interior region bounded by the curve and a noncompact (infinitely large) exterior region lying outside of the curve. (The Riemann mapping theorem just discussed concerns the conformal mapping of a region bounded by a Jordan curve onto a disk.) Jordan had previously shown that the area of a "rectifiable curve"—one of finite length—is always zero. But the question of whether the area of a general Jordan curve, including those of infinite length, had to be zero remained unanswered until Osgood came along. His paper answered that question in the negative by providing an example of a Jordan curve with positive area.[26] The idea here is along the lines of a curve that can fill up a region of space (such as the self-intersecting curve, discovered in 1890 by the Italian mathematician Giuseppe Peano, which passes through every point of a unit square).

More than just a novelty, the particular curve that Osgood constructed helped mathematicians revisit the question of what area is. The curve is also an example of a fractal, and the mathematician Benoit

Mandelbrot considered this example—and other curves of positive area subsequently constructed—in the course of developing his influential theories of fractal geometry.

In 1907, Osgood published the first volume of what would ultimately become a three-volume treatise on function theory, with the second volume coming out in 1924 and the third in 1932. The book was published in German, because Osgood believed the book would be more widely read—as well as more widely understood—than if it had appeared in English under an American imprint. Spanning most of Osgood's career, these three volumes can be viewed, in some sense, as his life's work. "The work, as a whole, is one of America's greatest contributions to the development of mathematics," Brown University mathematician Raymond C. Archibald claimed in 1938.[27] Osgood's text, stated Birkhoff, "for decades served as an unrivalled treatise on the subject,"[28] providing "a large part of the present mathematical world with its fundamental training in the field."[29]

Osgood once told Koopman that he was a physicist at heart, "that if during his student days the career of physics had offered a more deeply mathematical and less disproportionately experimental form, he might well have made it his. And there was ever present in his mind the notion of mathematics finding one of its deepest justifications in serving as an instrument whereby the mind of man can penetrate into the mysteries of nature." In trying to convey this idea to a student, Koopman added, "Osgood spared no patient effort in bringing him to see what mathematics truly is: the powerful agent of illumination of the physical world, and the gem of human reason."[30]

Bôcher also had a longstanding interest in mathematical physics and took on a problem of a similar nature, involving potential functions, during the three years he studied in Göttingen under Klein, pursuing work of this nature throughout his career. The potential function, as the name implies, can be used to describe, in precise mathematical terms, the electrostatic potential energy due to the distribution of charged particles, as well as the gravitational potential energy due to the distribution of mass, among other examples. As such, studies of the potential function have both an obvious and critical bearing on physics. But it is also an important part of pure mathematics, especially in analysis, which turned out to be the main area of research for both Bôcher and Osgood.

Klein initially steered Bôcher toward a problem involving the potential function, though Bôcher carried this work far on his own, finding solutions to partial differential equations using series methods, that is, by representing the potential function as the sum of an infinite series (also called a power series). Bôcher received a prize for the effort, which had been established by Klein for progress on this front, while also receiving his Ph.D. (The thesis was titled "Development of the Potential Function into Series.") He returned to Harvard in the fall of 1891 to serve as a mathematics instructor, as Osgood had done a year earlier. And like Osgood, Bôcher also brought a German bride back home with him to Cambridge.

Burdened by a heavy teaching load, including twelve hours of lectures each week, Bôcher was frustrated by the fact that, through Christmas of his first semester, he barely had any time to pursue his own research, though he tried to work over the holidays and any other time he could find a spare moment. He decided to expand his dissertation into a book on potential theory, which he began working seriously on during spring break. He continued to work on it for two years, obtaining new results that went well beyond the thesis. Klein was impressed with the outcome of that effort, and the book was published in Germany in 1894. Why did Bôcher, like Osgood, choose to publish in Germany rather than in the United States? Willie Sutton once said that he robbed banks because "that's where the money is." Bôcher and Osgood published their early books in Germany for similar reasons: that is where the mathematicians were—a larger potential audience because, in the 1890s, there simply were not that many people in the United States who would have read the book, let alone made sense of it. But Bôcher hoped to impress more than Felix Klein through his efforts. By turning out quality publications in a timely manner, as he and Osgood were doing, he hoped to show European mathematicians what their American counterparts were capable of.

Despite the dearth of time set aside for research, Bôcher somehow managed to produce important papers at an impressive rate. In 1892, for instance, he published five papers, and throughout the course of his unfortunately brief career he published approximately one hundred papers, all told, on differential equations, algebra, and other topics, in addition to his other textbooks on algebra, integral equations, analytic geometry, and trigonometry. Much of his research revolved around potential theory and Laplace's equation, a second-order partial differential equation that lies at the heart of function theory. "Here one stands in natural contact with mathematical physics, the theory of linear differential equations, both total and partial, the theory of functions of a complex variable, and thus directly or indirectly with a great part of mathematics," Birkhoff wrote of Bôcher's contributions.[31] Just as Bôcher started out working on potential theory, first during his thesis work under Klein and later for his first book, "nearly all of Bôcher's later work centered around the potential equation [too], which is indeed a focal point for a great part of analysis, real or complex," Birkhoff noted.[32]

In 1900, Bôcher wrote a notable paper concerning linear differential equations, which happened to be one of his specialties. In this paper, he demonstrated a new way to deal with functions that have discontinuities or "singular" points. Unlike a continuous function, a discontinuous function can have gaps and jumps. Singularities are places where functions are not "well behaved" or are otherwise poorly defined. For example, $f(x) = 1/x$ has a singularity at $x = 0$, where the function goes to infinity. At the time, standard methods, such as differentiation, were ill equipped to deal with this type of situation. But Bôcher found another way to proceed. He developed techniques for finding solutions around a singular point by making a series of approximations.

Bôcher's paper appeared in the first issue of a new journal, *Transactions of the American Mathematical Society*, to which he and other American mathematicians were eager to submit the best examples of their work. (Osgood, for instance, published his proof of the Riemann mapping theorem in the *Transactions*, also in 1900.) Bôcher was a founder of the journal and followed E. H. Moore as its second editor in chief, a position he held for five years.

A theorem that Bôcher proved in 1903, subsequently dubbed Bôcher's theorem, lies in the realm of harmonic analysis, which is crucial for understanding wave motion. The theorem relates to finding a solution to Laplace's equation for a function that is not well defined at the origin—or center of the coordinate system—which makes it especially difficult to work with. Bôcher showed that this hard-to-manage function could instead be rewritten as the sum of a function that is well defined at the origin—and thus easier to handle—plus a constant times another well-known function called Green's function. A year later, Bôcher proved another important theorem—which, to add to the confusion, is also referred to as Bôcher's theorem. The second Bôcher's theorem lies in the area of complex analysis, a theory of functions involving complex (as opposed to real-number) variables.

Bôcher returned to the subject of discontinuous functions in a 1906 paper on Fourier series—a technique developed by the French mathematician Joseph Fourier for representing a periodic function as an infinite sum of simple sine and cosine curves. The Yale physicist J. Willard Gibbs had previously remarked on a curious phenomenon involving successive curves that approximate a discontinuous function near the point of discontinuity. In particular, Gibbs found that the approximation by Fourier series exhibits large oscillations in the neighborhood of the discontinuity, with the Fourier sums consistently reaching a higher maximum value than the function itself. Bôcher dubbed this the "Gibbs phenomenon" while providing the first rigorous mathematical description of the effect. "His unusual modesty led Bôcher more than once to conceal a new result under the guise of an apparently expository article," Birkhoff noted. "This happened, for instance, in his beautiful exposition of the elementary theory of Fourier series, where the first detailed study of Gibbs' phenomenon was run into the body of the text almost without comment."[33]

This tendency was indeed characteristic of Bôcher, notes Joseph D. Zund, a mathematician formerly at New Mexico State University. "It is difficult to adequately appreciate Bôcher's influence on his time because so much of his work was devoted to perfecting and polishing material rather than to producing striking new results that would bear his name," Zund writes. "Nevertheless, his instinct and sense of what was important was impressive, and much of his work became commonplace in knowledge, although his authorship was largely forgotten."[34]

Bôcher's work was "basic," rather than "spectacular," according to Osgood, who meant basic in the best possible sense. "Creating for the mere sake of creating was a thing in which Bôcher did not believe ... Productive scholarship meant to him something larger and grander than the discovery of a new truth ... Productive scholarship meant to him a deeper insight into mathematics." Bôcher was keenly aware that most new theorems, which do not measure up to the standard of usefulness, have little if any value. He used his notions of productive scholarship to guide, and prioritize, his academic efforts. "I never knew a time in Bôcher's life in which he was not working on the most important problem of those among which he could reasonably expect to make an advance," Osgood said. As such, he was constantly striving "for the highest and purest beauty in mathematics."[35]

Though Bôcher "never occupied himself with an unimportant problem," as Archibald put it, he was still extremely prolific.[36] "In amount and quality his production exceeds that of any American mathematician of earlier date in the field of pure mathematics," Birkhoff wrote of Bôcher in 1919.[37]

Like Peirce, Bôcher was interested in philosophical questions concerning the nature of mathematics, and his general approach to mathematics was quite different from that of Osgood. Whereas Osgood was systematic, comprehensive, and fastidious about details, Bôcher believed that intuition, inspiration, and spontaneity had important roles to play alongside mathematical rigor. He explored these ideas—as well as the general question of what exactly mathematics is—at the International Congress of Arts and Science held in St. Louis in 1904. The old notion that mathematics is the science of quantity or numbers, he said, "has pretty well passed away among those mathematicians who have given any thought to the [matter]." One could try to characterize the subject by the methods employed, as Peirce did in calling mathematics "the science which draws necessary conclusions." By that definition, Bôcher said, "there is a mathematical element involved in every enquiry in which exact reasoning is used." The approach has two drawbacks, he added, the first being that "the idea of drawing necessary conclusions is a slightly vague and shifting one. Secondly that it lays exclusive stress on the rigorous logical element in mathematics and ignores the intuitional."[38]

A different way of characterizing mathematics—the approach favored by the British mathematician Alfred Kempe—is to focus not on the methods of mathematical investigation but rather on the objects of said investigation to see what those various objects might have in common. One drawback of Kempe's definition, which holds that mathematics is not strictly a deductive endeavor, is that it is too broad, Bôcher maintains: "It would include, for instance, the determination by experimental methods of what pairs of chemical compounds of the known elements react on one another when mixed under given conditions."[39]

Bôcher saw merit in combining Peirce's and Kempe's definitions, in keeping with the view of British philosopher Bertrand Russell (who spent several months at Harvard as a lecturer in 1914). Bôcher argued that focusing exclusively on the deductive side of mathematics represents only "the dry bones of science . . . Instead of inviting you to a feast, I have merely shown you the empty dishes and explained how the feast would be served if only the dishes were filled." He viewed mathematics as more art than science. "Rigorous deductive reasoning on the part of the mathematician may be likened here to technical skill [in a] painter ... Yet these qualities, fundamental though they are, do not make a painter or a mathematician worthy of the name, nor indeed are they the most important factors in the case." To Bôcher other qualities, such as imagination, intuition, and experiment of the nonlaboratory sort, were indispensable to the whole enterprise, as well as being integral to his own work.[40]

Osgood, on the other hand, stressed thoroughness, while placing less trust in artistic impulses. "Osgood's theorems ... were notable for their sharpness and ... rigor," commented Harvard mathematician Garrett Birkhoff, George Birkhoff's son. "Bôcher, on the other hand, was intuitive, brilliant, and fluent."[41]

These contrasts in mathematical approach and personality were also accompanied by pronounced differences in teaching and lecturing style. Osgood's exposition, recalled Walsh, "though not always thoroughly transparent, was accurate, rigorous, and stimulating."[42] Bôcher's elocution style stood in marked contrast. "His lectures were so lucid, the difficulties of subjects were not perhaps as effectively assimilated by students as they would have been under a poorer teacher," Archibald wrote.[43] In the journal *Science*, Bôcher was called the "preeminent" lecturer among American mathematicians.[44]

The same, alas, could not be said about Osgood, who was sometimes referred to by Harvard students as "Foggy Bill." Joseph Doob, a mathematician who received a bachelor's degree from Harvard in 1930 and a Ph.D. in 1932, reflected years later on the sophomore calculus class he took with Osgood, who taught the course using his own textbook.

> I did not suspect that he was an internationally famous mathematician and, of course, I had no idea of mathematical research, publication in research journals, or what it took to be a university professor. Osgood was a large, bearded portly gentleman who took his life and mathematics very seriously and walked up and down in front of the blackboard making ponderous statements. After a few weeks of his class I appealed to my adviser Marshall Stone to get me into a calculus section with a more lively teacher. Of course Stone did not waste sympathy on a student who complained that a teacher got on his nerves, and he advised me that if I found sophomore calculus too slow, I should take junior calculus at the same time![45]

Norbert Wiener found Osgood's German affectations to be annoying in the extreme. Wiener had this to say about the Harvard Mathematical Society meetings, which he regularly attended:

> The professors sat in the front row and condescended to the students in a graciously Olympian manner. Perhaps the most conspicuous figure was W. F. Osgood, with his bald oval head and his bushy divaricating beard, spearing his cigar with his knife after the pattern of Felix Klein, and holding it with a pose of conscious preciosity.
>
> Osgood typified Harvard mathematics in my mind. Like so many American scholars who had visited Germany at the beginning of the century, he had come home with a German wife and German mores. (Let me add that there is much to be said for marrying German wives—I am happy to have done so myself . . .) His admiration of all things German led him to write his book on the theory of functions in an almost correct German.[46]

To Wiener, Bôcher was "another representative of the German period of American mathematical education," but he, unlike his colleague Osgood, "was free from easily traceable mannerisms."[47]

Even colleagues who paid homage to Osgood in the journal *Science*— Julian Coolidge, George Birkhoff, and Edwin Kemble—conceded that Osgood's infatuation with Germany sometimes went too far. "He adopted the German Weltanschauung [world view] to an extent that became somewhat embarrassing during the First World War," they wrote, although, fortunately, "he saw matters in a different light" during World War II.[48]

As teachers, Bôcher and Osgood were "of equal but opposite excellence," wrote Coolidge. "Osgood acquired skill in teaching by the same process that brought him scientific eminence: conscientious effort and high ideals. He found out by experiment what were the most important things to teach, and what was the best way to teach them. Nothing was left to chance or the inspiration of the moment. When he finished there were no loose ends." Bôcher, by comparison, "never made an appreciable effort to be clear or interesting. His teaching was clear because his mental processes were like crystal; his teaching was interesting because he cared about interesting things."[49]

For a combination of reasons, some of which have already been delineated, Bôcher attracted far more graduate students than did Osgood. While Bôcher supervised seventeen Ph.D. theses in twenty-four years, Osgood supervised just four theses in forty-three years. As a thesis supervisor, Bôcher was "extraordinarily successful in discerning important topics with range and depth, yet still within the powers of students at that stage of advancement," Osgood notes. "He was skillful in his guidance, not giving help that hampered the student's initiative or resourcefulness, nor neglecting him when he was ready to profit from a conference."[50]

Osgood, however, was neither as skillful nor as successful in these same areas. He had achieved clarity, as Coolidge had said, only through considerable effort, though he did maintain that lucidity was overrated. "The clearer the teacher makes it, the worse it is for you," he told his students, according to the memoirs of the mathematician Angus E. Taylor. "You must work things out for yourself and make the ideas your own." As for how to master a difficult concept, Osgood suggested the following strategy: "Read it over a couple of times and sleep on it. Then, when you think you've grasped the situation, tell yourself the whole story in your own words, perhaps by writing it on a scrap of paper while on the subway to Boston."[51]

Simply put, not many students were inspired by Osgood in their research work. The reason for that, according to Birkhoff, is not hard to find: "His method of attacking a problem was to make an exceedingly careful and systematic exploration of detail; it was thus that his own creative ideas came to him after arduous effort. It was natural then for Osgood to suggest to the prospective student desirous of working with him that he first make a careful preliminary routine survey of the field. But the average student was discouraged at the very outset by the unaccustomed labor, which was thus required without apparent reward."[52]

David Widder, a Harvard math Ph.D. who later taught in the department, also found Osgood's style less than exhilarating. "I recall that he gave us good advice, ignored by most, on how to prepare a paper. You were to fold it down the middle, put a first draft on the right, corrections on the left."[53] Widder chose to do his Ph.D. work under Osgood's younger colleague, George David Birkhoff.

Osgood might have supervised one other Ph.D. student had he not been so keen on having a broad command of every last fact before proceeding. Joseph Walsh asked him whether he might supervise his thesis on a subject connected to the expansion of analytic functions. Osgood threw up his hands, Walsh said, insisting that he knew "nothing about it."[54] Walsh got Birkhoff instead to oversee his dissertation work, just as Widder had done.

Fellow faculty members identified these reasons for Bôcher's success in guiding graduate students in their research. "It is not difficult in science to find important problems which cannot be solved, or unimportant ones which can be," his Harvard colleagues wrote.

> Bôcher was successful in discovering subjects on which the advanced student could work with a reasonable prospect of securing results of value. He did not foster research by excessive praise, and his students sometimes felt that he was unappreciative. But a scientific contribution of real merit never failed to secure his attention, and he had infinite patience in helping the student who was really making progress to develop his ideas, to see that which was new in its true perspective, and to put his results into clear and accurate language.[55]

Bôcher brought those same attributes to his responsibilities as editor in chief of the *Annals of Mathematics*—a role he served from 1908 to 1914 (not counting 1910), during which time he maintained a high standard, according to a tribute paid to him in the *Transactions of the American Mathematical Society* (to which Bôcher had been a frequent contributor). "No contribution received his approval merely because of the reputation of the author, but if he saw real merit in the work of an inexperienced writer, he was liberal in encouragement and suggestion."[56]

The two men, Bôcher and Osgood, whose careers followed such parallel tracks over the years nevertheless had rather sharp divergences in personality. Osgood was described as "warm and tender" by one colleague—words that were rarely applied to Bôcher. To students, as Koopman recounted, "Osgood was a happy combination of the wise master and the cordial friend."[57] Or, as fellow faculty members put it, "He was unwearied in his acts of kindness to individual students, and he treated all with an old-fashioned courtesy which sprang from his deep love for his fellow man."[58]

Osgood's favorite forms of recreation were said to be touring in his motorcar, though he also enjoyed the occasional game of golf and tennis, as well as smoking cigars. "For the latter, he smoked until little of the cigar was left, then inserted the small blade of a penknife so as to have a convenient way to continue," explained Walsh, affectionately describing a pastime that Wiener, by contrast, found particularly irksome.[59]

Bôcher was more reserved and private than Osgood, who considered himself to be reserved. "[Bôcher] sought relaxation from scientific labor in literature [Shakespeare, the French and German classics, and biography], philosophy, and music, rather than in social gatherings," reported an obituary in the journal *Science*.[60] "He never gave way to enthusiasm," wrote Archibald. "He was a puritan, and with the virtues he had also the faults of the puritan. There was no place in his world for human weakness; he respected only results. These same stern standards he applied equally to himself."[61]

"To many a man who came into personal relations with him in his profession, he seemed cold and unsympathetic," Osgood explained. "What the stranger, however, too often failed to observe was that Bôcher applied the same stern standards to himself. Why should others expect to fare better?" Bôcher never showed in his interactions with others any hint of "the enthusiasm of just doing things in mathematics—the joy of living, so to speak—[that] gives to one's mental work a momentum which carries it over the obstacles of disappointment and discouragement." His nature was, at all times, much more withdrawn. "He would not talk on personal matters relating to himself, and this disinclination extended even to his scientific work."[62]

Bôcher suffered from poor health during many of his later years, and sadly, his life was cut short by illness. He died in 1918 at the age of fifty-one, while still in the prime of his career. In 1923, the American Mathematical Society established the Bôcher Memorial Prize in his honor to recognize notable research in the field of analysis to which Bôcher himself had contributed greatly.

Osgood's tenure at Harvard was also cut short, though for entirely different reasons. A scandal led to his retirement in 1933 after he married Celeste Phelps Morse—the former wife of Harvard mathematics professor Marston Morse—in 1932. Morse, who was twenty-eight years younger than Osgood, was shocked that his former wife would take up with an older man sporting a full white beard. The development may have been sparked by a conversation Celeste Morse once had with Osgood about her marital difficulties. Osgood ended up divorcing his German wife, just as Celeste divorced Morse, clearing the way for their eventual union.

In August 1932, some time after the scandal had broken out, Harvard president Abbott Lawrence Lowell received a letter from the physician Roger Lee, which said: "I quite agree that in the case of Professor Osgood it would be well to ask him to retire on account of age as soon as it is decent."[63] Lowell wrote to Osgood two months later asking for his resignation at the end of the academic year. Lowell cited a provision in the university's bylaws whereby "any professor shall retire on his pension at sixty-six if requested to do so." Osgood was sixty-eight years old at the time of this letter and sixty-nine when he retired in 1933. "It is sad to have to write this to a colleague who has served so long and earnestly," Lowell wrote.[64]

Osgood and his new wife moved to China; he lectured for two years at Peking University. Two textbooks based on the lectures he delivered in Peking were later published: *Functions of Real Variables* (1936) and *Functions of a Complex Variable* (1936). After two years in China, Osgood and his wife moved back to the Cambridge area, settling in the adjacent town of Belmont, Massachusetts, where Osgood reportedly enjoyed his retirement years. He died in 1943 at the age of seventy-nine.

Bôcher and Osgood had much to be proud of: They helped put the mathematical discipline of analysis on a solid footing in the United States. They left behind a math department that not only was among America's best but also was competitive with the strongest departments across the Atlantic. Their leadership position at Harvard was taken over by a former student, George David Birkhoff, who had learned from them as an undergraduate. Birkhoff earned his Ph.D. under Moore at the University of Chicago, but Bôcher maintained contact with his former student, continuing the correspondence through Birkhoff's teaching stints at the University of Wisconsin and Princeton. Harvard offered him a job in 1910, but Birkhoff turned it down when Princeton offered him a promotion and raise. Birkhoff had second thoughts a year or so later, asking Bôcher whether a position might still be available for him. A deal was soon worked out that brought Birkhoff to Harvard in 1912—the same year that Bôcher was asked to deliver a plenary address at the International Congress of Mathematicians in Cambridge, England. It was also the same year, not coincidentally, that Harvard made "the transition from primary emphasis on mathematical education . . . to primary emphasis on research."[65]

Birkhoff remained at Harvard for the rest of his life. With that appointment, which placed him alongside his former teachers, Bôcher and Osgood, "Harvard was the strongest department in the United States," writes Batterson. It was also the strongest mathematical faculty ever assembled in the country up to that time, he adds, poised to become a power on the world stage. "Given the 1890 state of American mathematics, the rise of Harvard was remarkable."[66]

Birkhoff represented the next generation of American scholars, trained entirely in this country. While still in his twenties, he became known throughout the world for his mathematical abilities, showing, in case there was need of further demonstration, that one need not go to Europe to secure a world-class education. He and other top-notch mathematicians were graduating from American universities, well equipped to lead their fields and their departments in the near future. A domestic mathematics infrastructure was thus emerging, thereby satisfying a dream that Benjamin Peirce long held but did not see realized in his lifetime.

The rise of Birkhoff to a position of dominance in Harvard mathematics, and in U.S. mathematics in general, was a signal, of sorts, that the transformation that began in the late 1800s had now reached fruition. "No longer did college mathematics mean merely arithmetic, trigonometry, the rudiments of algebra and geometry, and a smattering of calculus," writes Parshall. "No longer were American students essentially forced to travel to the great universities of Europe if they wished to study the modern advances in mathematics seriously."[67] As far as American mathematicians were concerned, you could go home again—despite Thomas Wolfe's words to the contrary. And more important—perhaps to some parents' lament—you did not have to leave home in the first place. Birkhoff, along with other standard-bearers of mathematics' new generation, would go on to prove important theorems and make many remarkable contributions—a story that will be taken up next.

3

THE DYNAMICAL PRESENCE OF GEORGE DAVID BIRKHOFF

Within a year of joining the Harvard faculty in 1912, George David Birkhoff had knocked off a celebrated problem, posed by Henri Poincaré, that would earn him international fame. Birkhoff solved a version of the three-body problem, which involves describing motions in a system of three gravitating objects such as the sun, Earth, and the moon. The eminent mathematician Poincaré had framed the classic problem in a new way. And since his health was failing at the time, he had put the problem out to the rest of the world, hoping that someone might find a way through it. That someone turned out to be Birkhoff in what was, for him, the first of many high-profile triumphs.

New assistant professors do not normally make such a stir, but Birkhoff was not your typical assistant professor. For many who knew him, his feat—occurring in just his first year at Harvard—was hardly surprising. And odds are that Birkhoff himself was not surprised either, since he was driven to excel from an early age and acknowledged few, if any, limits on what he might ultimately achieve. "He always regarded himself as competing with the greatest mathematicians of all time," said his son Garrett, a mathematician who also spent his career on the Harvard faculty.[1]

The son of a physician, George Birkhoff was born in Michigan in 1884 to parents of Dutch descent. He grew up mostly in Chicago, the product of an exemplary education. Birkhoff stood out from his predecessors, and many contemporaries as well, by getting his mathematics training strictly in the United States, rather than going abroad. For high school, he attended the Lewis Institute in Chicago (which later became part of the Illinois Institute of Technology), which is where he acquired his love of mathematics. By the age of fifteen, he was solving problems posed in the *American Mathematical Monthly*. One of the problems he solved sounded rather elementary: "Prove that if two angle bisectors of a triangle are [of equal length], the triangle is isosceles." Yet the exercise had actually "caused great difficulty ... to a number of mathematicians in France, England, and the United States, including several leading ones," according to Harry Vandiver, a Birkhoff contemporary who served on the University of Texas mathematics faculty for forty-two years prior to his death in 1973.[2]

Vandiver—a self-taught mathematician who dropped out of high school and never went to college—began corresponding with Birkhoff when both were teenagers, solving problems posted in the *Monthly*. They eventually collaborated on dozens of papers together, including a paper on number theory that was published in the *Annals of Mathematics* while Birkhoff was still an undergraduate.

As a teenager, Birkhoff also established a previously unknown link between an important problem in number theory and the even more important Fermat's last theorem. He told Vandiver of his interest in Pierre de Fermat's classic problem, which asserts that no positive integers a, b, and c can be found such that $a^n + b^n = c^n$, where n is a positive integer greater than 2. Although the problem can be stated quite succinctly, and simply, it remained unsolved for nearly 360 years after Pierre de Fermat first presented it in 1637. Vandiver became interested in the problem, too, and ended up working on Fermat's last theorem throughout his career. In 1931, he won the first Cole Prize in number theory ever granted, mainly for his efforts to solve the problem. The prize was named after Frank Nelson Cole, the former Harvard Ph.D. recipient and lecturer who had inspired Maxime Bôcher and William Fogg Osgood as undergraduates, before he joined the faculties of the University of Michigan and Columbia.

In a letter to Vandiver written decades earlier, a youthful Birkhoff had vowed to his friend: "We will solve Fermat's problem and then some!"[3] Birkhoff did not make good on his first promise, but more than made up for it on the second.

By the time Birkhoff entered the University of Chicago as a freshman in 1902, Garrett wrote, "he had already begun his career as a research mathematician."[4] Birkhoff moved around a bit in academia, spending two years as an undergraduate at the University of Chicago before spending two years at Harvard, where he earned his bachelor's and master's degrees. The move might have been motivated, in part, by his uncle Garrett Droppers, who had previously attended Harvard before going on to become a prominent economist. It is also likely that the switch fit in with Birkhoff's goal of exposing himself to the best teachers and most stimulating environments available. Although he had excellent instructors at Chicago—including E. H. Moore, Oskar Bolza, and Heinrich Maschke—after two years he might have felt the need to broaden his horizons and, in the process, spark some new ideas. Bôcher, whose research interests were closer to Birkhoff's than were Moore's, helped fill that role at Harvard, for which Birkhoff remained grateful. He later thanked Bôcher "for his suggestions, for his remarkable critical insight, and his unfailing interest in the often-crude mathematical ideas, which I presented."[5]

Birkhoff, however, did not hit it off so well with Bôcher's colleague Osgood. "He felt that Osgood was more or less a martinet, who had the German teacher-dominating-student approach," Garrett explained. "Bôcher was much more flexible and informal, and was generally considered a more inspiring, if less systematic, teacher."[6]

Mathematical research was important to Birkhoff throughout his undergraduate years, although that interest was sparked even before he entered college. From the very beginning, he was solving problems—the harder the better—and publishing the results in various journals. He maintained this practice, not surprisingly, as a graduate student of Moore's at Chicago and continued to publish—upward of one hundred journal articles and numerous textbooks on dynamical systems, graph theory, ergodic theory, general relativity, "aesthetic measure," and other topics—over the course of his career. "Birkhoff was uncompromising in his appraisal of mathematics—by the test of originality and relevance," said the mathematician Marston Morse, a Ph.D. student of Birkhoff's who went on to a successful career at Harvard and the Institute for Advanced Study in Princeton, New Jersey. Birkhoff applied these same standards to his own work. For him, Morse added, "the systematic organization or exposition of a mathematical theory was always secondary in importance to its discovery."[7]

Never one to settle for mere proficiency, Birkhoff was intent on achieving mathematical distinction of the highest order. This desire can be seen in preparations for his Ph.D. orals, where he set the goal of trying "to master essentially all the mathematics that was known at the time," Garrett recalls. "And [he] made a good stab at achieving this ambition!"[8]

Although Moore and Bôcher had undeniable impacts on Birkhoff, another scholar, Poincaré, may have exerted an even greater influence, even though the two of them never met. "Poincaré was Birkhoff's true teacher," Morse maintained. "In the domains of analysis," which involves the application of advanced forms of calculus, "Birkhoff wholeheartedly took over the techniques and problems of Poincaré and carried on."[9]

In particular, Birkhoff shared Poincaré's interest in celestial mechanics, learning from the master by poring over Poincaré's acclaimed book *New Methods of Celestial Mechanics*—a successor to Laplace's venerated work in the same area. "In a very literal sense, Birkhoff took up the leadership in this field at the point where Poincaré laid it down," stated Oswald Veblen, a lifelong friend of Birkhoff's who overlapped with him at the University of Chicago and Princeton.[10]

Birkhoff's Ph.D. work was in this same area of "dynamics": the use of differential equations to describe the motions of objects in complicated systems, where the motion of one object affects the motion of others. In this way, Birkhoff was following in the tradition of other luminaries with Harvard ties—Nathaniel Bowditch, Benjamin Peirce, Simon Newcomb, and George William Hill—all of whom either worked at the university proper or worked or studied with Peirce at some point in their careers. Of course, Birkhoff did more than just continue this tradition. While Poincaré justifiably impressed him, Birkhoff developed new tools that went beyond the French master and other predecessors, thereby leading to new, previously inaccessible results. In this way, he helped usher in the modern theory of dynamical systems. And like Poincaré, he divided his time between pure and applied mathematics.

Much of Birkhoff's work was motivated by an overriding goal: to reduce the most general dynamical system to a "normal form" from which a full and detailed characterization could be drawn. The trick here, in a sense, is finding the optimal way of framing a problem: a proper coordinate system that can enable one to complete calculations that might not be possible otherwise. An ellipse, for example, is ordinarily described by a rather complicated and unwieldy equation:

$$x^2/a^2 + xy/b + y^2/c^2 + ex + fy + g = 0,$$

where a, b, c, e, f, and g are constants. However, by choosing suitable coordinates and centering the ellipse at the origin of the x-y plane, one can put the ellipse into its normal form, $x^2/a^2 + y^2/b^2 = 1$, which is much easier to look at and work with. Birkhoff used the same approach in dealing with dynamical systems, trying to frame the situation with good coordinates, thereby "normalizing" it so that one can complete calculations that might not otherwise be done.

He achieved his first high-profile success, as mentioned earlier, on the "restricted" three-body problem posed by Poincaré. The challenge in problems of this sort is to take the initial conditions—the position, velocity, and mass of three separate, though interacting, objects—and then predict their motions into the future, with one goal being to find stable, periodic solutions. The three-body problem has long been regarded as "the most celebrated of all dynamical problems."[11] It has yet to be solved in complete generality, and such a solution may not be possible owing to the presence of chaotic phenomena and the fact that the motions are described by differential equations that cannot be integrated.

The two-body problem—which concerns the motions of two objects that interact only with each other—can, by contrast, be solved in a straightforward fashion using elementary functions that are completely integrable. The situation is much simpler, in part, because two objects (of the sort discussed in such cases) inevitably lie in the same plane. About three hundred years ago, Isaac Newton solved a version of the two-body problem when two objects interact through gravity—a force whose strength varies inversely with the square of the distance between them. Newton's calculations were a great triumph, providing mathematical proof of the laws of planetary motion that Johannes Kepler had discovered a century earlier (based on the observations of Tycho Brahe) to describe the paths of two bodies in orbit around each other.

But in a system with three objects, motions need not be confined to a plane, which results in a much more complicated nonlinear problem that can be exceedingly difficult to solve. Newton, once again, was among the first to take up this problem, studying the system comprising Earth, the sun, and the moon. He first calculated the moon's orbit around Earth but had great difficulty computing the sun's effect on the moon's motions, telling a fellow astronomer that "his head never ached but with his studies on the moon."[12]

One way to make progress on such a problem is to consider a simpler, special case—the so-called restricted case. In this situation, two bodies move in tandem owing to their gravitational attraction, but there is a third body as well, whose mass can safely be ignored. Although the mass of the third body is considered negligible insofar as the other objects are concerned, this is still a three-body problem, because the whole point of this exercise is to figure out how the motions of the two massive bodies affect the motion of the essentially "massless" third.

Poincaré had been pondering the three-body problem for decades, noting that it "is of such importance in astronomy, and is at the same time so difficult, that all efforts of geometers have long been directed toward it."[13] His thinking in this area was strongly influenced by the American George William Hill, whose 1877 paper involving interactions of the sun and moon presented the first new periodic solutions to the three-body problem in more than a century. During a visit to the United States, Poincaré said to Hill: "You are the one man I came to America to see."[14]

Poincaré achieved his own success in this area as well, winning an international prize in 1890 (in honor of the sixtieth birthday of Oscar II, the king of both Sweden and Norway) for a paper he wrote about the three-body problem for *Acta Mathematica*. Some years later, Birkhoff said that the journal had published many important articles, "but perhaps none of larger scientific importance than this one."[15]

Despite Poincaré's acknowledged genius, and the effort he had applied in this direction, he was stumped by the three-body problem—along with everyone else who had grappled with it since Newton first provided the tools that enabled people to at least consider tackling it. Poincaré fell ill in 1911. Sensing that he may not have long to live, he secured permission from the editor of the *Circolo Matematico di Palermo* to publish an incomplete paper on this subject, "in which I would give an exposition of the goal that I have pursued, the problem that I have set for myself, and the result of the efforts that I have made to solve it." Although Poincaré realized that publishing an unfinished and, to some extent, unsuccessful paper "would be somewhat unprecedented," he suggested that such a paper might still be useful by "putting researchers on a new and unexplored path" that is "full of promise, despite the disappointment they have caused me."[16] The paper, which appeared in 1912, cast the restricted three-body problem in a new light, offering a promising means of attack. Poincaré died on July 17, 1912, a few weeks after the paper came out. His hope, in having his thoughts on the "last geometric theorem" (as it came to be known) published prior to its proof, was that some other mathematician might follow the path he had outlined to success.

Birkhoff did just that, completing the task within three months of Poincaré's death, thereby sealing his own reputation in mathematics. Although Birkhoff solved the problem rather quickly, he did have to exert himself, later telling a student (Marshall Stone) that he lost thirty pounds working out the proof.[17]

Birkhoff's 1913 paper "Proof of Poincaré's Geometric Problem" was succinct (a mere nine pages long) yet elegant. It stated the problem as follows: "Let us suppose that a continuous, one-to-one transformation T takes the ring R, formed by concentric circles C_a and C_b of radius a and b, respectively ($a \geq b \geq 0$), into itself in such a way as to advance the points of C_a in a positive sense and the points of C_b in the negative sense, and at the same time to preserve areas. Then there are at least two invariant points."[18]

The problem, as laid out by Poincaré, "represents one of the first instances of reducing an existence theorem in analysis to a topological fixed point problem," wrote the former Harvard mathematician Daniel Goroff.[19] To express the idea in somewhat simpler terms, let us start with a ring or annulus—a geometric figure encompassing two concentric circles and the area between them—or its topological equivalent, an open cylinder (like a can with its top and bottom removed). Now "transform" this shape by twisting one of the circles bounding the ring (or open cylinder) in, say, the clockwise direction and the other circle bounding the ring in the counterclockwise direction—without changing the area. For the moment, we are not saying anything about what is happening in the middle, just what is happening along the edges. But Poincaré asserted, nevertheless, that for an area-preserving transformation, which shifts one circle in one direction and the other circle in the other direction, there must be at least two points on the surface between these circles that do not move at all.

Poincaré maintained, further—based on a concept he introduced called the "index theorem"—that if there is one invariant point on a ring or annulus that has undergone an area-preserving transformation, there must be a second. Birkhoff used this argument in his proof by contradiction, or reductio ad absurdum, by showing that the assumption that there is no invariant point leads to an absurd consequence. Hence, that assumption must be incorrect, meaning that there is at least one invariant point. And if there is one such point, there must be two, as specified in Poincaré's theorem.

Birkhoff's 1913 paper proved the last geometric theorem in the simplest, two-dimensional case. He generalized the theorem to higher dimensions in a 1931 paper. Two decades later, the Russian and German mathematicians Andrey Kolmogorov, Vladimir Arnold, and Jurgen Moser made further dramatic extensions in what has come to be known as KAM theory.[20]

It is not obvious, however, what fixed points on an annulus have to do with the famous three-body problem in celestial mechanics. One way to think of it, explains Northwestern mathematician John Franks, is that "the annulus is like a two-dimensional slice of three-dimensional space." Objects, such as planets, are moving around in this three-dimensional space, which means they will occasionally poke through the two-dimensional annulus. If, after moving through this bigger space, they come back to the very same spot on the annulus, that would correspond to an invariant point. Returning, time and again, to such a fixed point would, in other words, imply a degree of stability, predictability, and periodicity: objects start at a particular place on the annulus and, after venturing into the big world, return to the same location on a regular basis.

But one fixed point on the annulus does not translate to a single body in the three-body system. The situation is sufficiently abstract, Franks adds, that a single invariant point instead corresponds to all three objects in the three-body system. Poincaré's theorem says, of course, that there would be not one but two invariant points. One of these points relates to the three-body problem; the second point is simply there as a topological consequence of the first.[21]

Once proved, the last geometric theorem offered a technique for finding an unlimited number of periodic solutions to the restricted three-body problem: by varying the size of the annulus, one could obtain different fixed points, with each point corresponding to different orbital periods for the bodies in the system.

In solving a version of the three-body problem—and showing how one might, in fact, obtain an infinite number of solutions—Birkhoff gained almost instant worldwide recognition. Speaking of his accomplishment, the renowned MIT mathematician Norbert Wiener claimed that "a star of the first magnitude had appeared on the firmament of the Harvard mathematics department . . . What was even more remarkable was that Birkhoff had done his work in the United States without the benefit of any foreign training whatever. Before 1912 it had been considered indispensable for any young American mathematician of promise to complete his training abroad. Birkhoff marks the beginning of the autonomous maturity of American mathematics."[22] Not only had Birkhoff never studied in Europe, he did not set foot on the continent until 1926—fourteen years after proving Poincaré's geometric theorem.

The year 1912 was a turning point for another reason, according to Garrett Birkhoff, who called it "a milestone marking the transition from primary emphasis on mathematical education at Harvard to primary emphasis on research." Within a year, William Elwood Byerly (Benjamin Peirce's first Ph.D. student) retired from the department, and B. O. Peirce died, marking "the end of Benjamin Peirce's influence on mathematics at Harvard," as Garrett put it, while also marking the beginning of Birkhoff's influence for decades to come.[23]

Far from resting on his laurels, Birkhoff kept up a brisk pace, turning out eight other papers in 1913 in addition to his Poincaré proof. "His problems were not necessarily chosen from among those problems which he could solve," Morse noted.[24] Garrett Birkhoff agreed that his father did not shy away from difficult problems, going out of his way to take on the most obstinate ones, attacking them, whenever possible, "with radically new ideas and methods."[25] That propensity, no doubt, led him to the four-color problem and his 1913 paper on the subject, "The Reducibility of Maps."[26]

For more than a century, the four-color problem stood as one of the most famous unsolved problems in topology—and one that many prominent mathematicians tried their hands at. Though easy to understand, the problem turned out to be surprisingly difficult to prove. It revolves around a simple question: How many colors are required for any map that could be drawn in a plane such that no two adjacent sections of the map have the same color? (As defined in this problem, the states of Colorado and New Mexico are "adjacent" and would thus be assigned different colors. On the other hand, Colorado and Arizona, or New Mexico and Utah, which simply share a corner in common, are not considered adjacent and could therefore have the same color.)

The four-color conjecture, as the name implies, holds that any map—no matter how large, convoluted, or squiggly—requires just four colors to ensure that any two neighbors with overlapping borders have different colors. The problem pertains to a map lying on a plane or on a sphere—on a surface of genus 0, which means that (unlike a doughnut) it has no holes.

Credit for the conjecture is given to Francis Guthrie, a mathematics graduate of the University College London and a former student of Augustus De Morgan, who posed it in 1852. After hearing of the problem, De Morgan passed it on to William Rowan Hamilton, the Irish mathematician of quaternion fame (discussed in Chapter 1). Hamilton replied a few days later, saying, "I am not likely to attempt your quaternion of colour very soon."[27]

Arthur Cayley, a prominent British mathematician, outlined the difficulties of the theorem in an 1879 paper. In that same year Alfred Kempe—a British lawyer as well as a mathematician—published a proof of the theorem, which earned him acceptance into the Royal Society two years later. In 1890, however, Percy John Heawood uncovered a fatal flaw in Kempe's alleged proof; Heawood proved conclusively that five colors are sufficient to guarantee that adjoining regions of the map have different colors, but he was unable to prove the more demanding case of four colors. Others who looked into the problem, but fell short of a proof, include Benjamin Peirce, his son Charles Sanders Peirce, and Oswald Veblen. Hermann Minkowski, the famed German mathematician who contributed greatly to relativity theory, thought he could solve the problem during a topology class. Although his students were likely entertained, Minkowski got frustrated when his proof went astray, and his planned classroom triumph ended in failure.

Birkhoff, who was always up for a challenge, could not resist the pull of the four-color problem, which became a longstanding hobby of his, if not an obsession. It has been reported, in a tale that is possibly apocryphal, that Birkhoff's wife asked another mathematician's wife, shortly after the latter's wedding, "Did your husband make you draw maps for him to color on your honeymoon, as mine did?"[28]

Vandiver, nevertheless, can attest to Birkhoff's fixation on the four-color problem: "I recall his telling me around 1915 that it was the only mathematical problem he had studied up to that time which had kept him awake at night." The problem interested him throughout his life, Vandiver said, although he was puzzled by his friend's preoccupation, especially since Birkhoff had told him he had not worked more on number theory because the main unsolved problems were so difficult that it would take an extraordinary amount of time to get anywhere with them—time he could put to better use by obtaining substantial results in other areas. "As time went on, however, it appeared that he was not entirely consistent in this," Vandiver added, "as he must have spent an inordinate amount of time on the four-color map problem."[29]

Birkhoff had not forgotten about the problem in 1942, when the topologist Solomon Lefschetz visited Harvard. Birkhoff asked him what was new at Princeton. Lefschetz said that a recent visitor to his math department had just solved the four-color problem. "I doubt it but if it's true, I'll go on my hands and knees from the railroad station to Fine Hall," Birkhoff promised. He never had to do that, as the alleged proof never materialized. Birkhoff was probably confident in his pledge, owing to the slew of fallacious proofs that had already been put forth.[30]

It would be wrong, however, to suggest that all of Birkhoff's efforts in this direction came to naught. Although his 1913 paper did not contain a proof, or attempt one, it did outline a strategy, involving the notion of "reducibility," that became central in most subsequent attempts to solve the problem. The chief difficulty stems from the fact that the theorem refers to all possible maps of all shapes and sizes; an argument must therefore work in every conceivable case. As the possibilities are literally endless, Birkhoff offered a way to reduce the number of regions in a map from an infinite number to something that is finite and, it is hoped, manageable. If you have a mathematically acceptable way of transforming a map with an infinite number of regions to one with a finite number, and if you can further show that four colors are sufficient for the smaller map, you can potentially show that four colors are sufficient in all cases—or so the argument goes.

Philip Franklin, a Ph.D. student of Veblen's who taught at Harvard, among other institutions, published a paper that incorporated Birkhoff's concept of reducibility, proving that only four colors are needed for any map consisting of twenty-five or fewer regions. The number of regions steadily increased over the years; in 1976, Jean Mayer, a mathematician whose principal academic appointments were in literature, would raise the "Birkhoff number"—the number of regions for which four colors suffices—to ninety-six. It is possible that no one went beyond ninety-six not because that represents an upper limit but rather because the four-color theorem was proved outright shortly after Mayer's contribution.[31] That proof, discussed later in this chapter, follows a path that had been forged by Birkhoff decades before.

While pursuing this problem, Birkhoff developed another tool that bears mention—the so-called chromatic polynomial, a function $p(x)$ that is equal to the number of ways of coloring a map in x colors. Chromatic polynomials are not only useful in attacking the four-color problem but are also important in the branch of mathematics known as graph theory, where they are still used today—a century after Birkhoff introduced the concept.

His work in this area was reminiscent of the attempts of the German mathematician Ernst Eduard Kummer to solve Fermat's last theorem in the mid-1800s. Although Kummer's proofs broke down, his approaches influenced subsequent work on the problem for years to come. These same ideas, which Kummer developed while working on the Fermat problem, became important in many branches of number theory. They also supported new developments in abstract algebra, while leading to the establishment of a new field called ring theory. One would be hard-pressed to label Kummer's efforts as failures, in light of all that has come from them, and the same applies to Birkhoff's efforts on the four-color problem.

Although he never fully abandoned the problem, Birkhoff never let it stand in the way of other projects, many of which yielded more immediate and tangible results. He published a paper in 1917, "Dynamical Systems with Two Degrees of Freedom,"[32] that he considered, according to Morse, to be "as good a piece of research as he would be likely to do."[33] Birkhoff was not alone in that opinion, apparently, as he won the 1923 Bôcher Prize—named in honor of his former mentor—for the work he did resulting in that paper. Birkhoff was the first recipient of this prize, which was, itself, the first prize of any kind ever issued by the American Mathematical Society, over which Bôcher had presided from 1909 to 1910.

Birkhoff's focus here was on dynamical systems in which motions are allowed in two independent directions or dimensions. He was able to simplify the picture dramatically, and bring some cohesion to the subject, by finding a standard template, or "normal form," that reduced an infinite number of possible motions into a finite number of general types. Or, as the English mathematician E. T. Whittaker put it, "he reduced all problems relating to such systems ... to the problem of determining the orbits of a particle constrained to move on a smooth surface which rotates about a fixed axis with uniform angular velocity."[34] By framing the possibilities in this way, explained Morse, Birkhoff came "close to realizing a complete, qualitative characterization" of such systems.[35]

The paper was also notable because Birkhoff introduced his new "minimax principle." To get a crude sense of what this is about, let us start with a simple two-dimensional surface, the sphere, and cover that surface with closed curves. There are an infinite number of ways of covering the sphere, especially when you consider that the curves can intersect and cover the surface more than once. Birkhoff showed that there is a function, f, that can be maximized to pick out the longest curve in a family of closed curves that sweep across the surface. The curve of maximum length is sometimes called the "generalized geodesic." The value of f tells us the length of this generalized geodesic. In addition, Birkhoff showed that among the infinite number of families, f will achieve its minimum value at one of the families. In this particular family, the length of the generalized geodesic will be smaller than the length of the generalized geodesic in the family of curves that can be generated by deforming the original family. The generalized geodesic picked out in this way can, in fact, be shown to be the geodesic as defined in classical geometry (such as the "great circle" of a sphere). Proving that a function in an infinite-dimensional space has a minimum—the "mini" part of Birkhoff's proof—is no easy task because, in general, a function in such a space may not have a minimum at all. But Birkhoff demonstrated how to obtain the so-called minimax geodesic by first taking the maximum among a family of curves and then finding the minimum by deforming the original family.

For that reason, Birkhoff's proof was a breakthrough, representing an early application of the calculus of variations to geometry. The principle served as the starting point for Morse theory, developed by Marston Morse, which applies the calculus of variations to topology. Morse theory (discussed in Chapter 4) was extremely influential, having a profound sway over differential topology for more than half a century.

In 1923, Birkhoff proved an important theorem in general relativity. In "Birkhoff's theorem," as it is called, he proved that the metric, or geometry, of the so-called Schwarzschild black hole is the unique solution to Einstein's equations for the gravitational field of a spherically symmetric object in a vacuum, with no other mass lying around it. A Schwarzschild black hole is a hypothetical entity: an idealized, "static" black hole with no charge or spin, which makes it easier for theorists to work with. The object is named after the German physicist Karl Schwarzschild, who hit upon this solution in 1915, a month after Einstein published his theory of general relativity. Black holes of this type have a standard geometry and can be distinguished from one another only by means of their mass.

Many researchers had tried to construct other spherically symmetric, vacuum solutions to Einstein's equations, but they always came back to the Schwarzschild solution. Birkhoff showed why that was the case. He was the first to prove that the Schwarzschild solution is the *only* solution—that you cannot get anything else. Birkhoff's uniqueness theorem, moreover, was much stronger than most. While uniqueness arguments typically rest upon many assumptions, Birkhoff's proof assumed nothing other than the spherical symmetry of a black hole.

In 1927, Birkhoff published a landmark book, *Dynamical Systems,* which built on Poincaré's ideas and generalized them. Birkhoff introduced new topological approaches, while broadening the scope of dynamics beyond celestial mechanics, so that the motions of all manner of objects could be considered, not just the orbits of planets and stars. This book, claims math historian David Aubin, "created a new branch of mathematics separate from its roots in celestial mechanics and making broad use of topology."[36]

Aubin considers *Dynamical Systems* a worthy successor to Poincaré's three-volume book on celestial mechanics, which George Darwin (Charles Darwin's son) called "for half a century to come . . . the mine from which humbler investigators will excavate their materials." Nevertheless, Aubin adds, "Poincaré's magisterial treatise contained much that was cumbersome to use, at times obscure, and at times—for those interested in general dynamics—unduly concerned with details of celestial mechanics. For Birkhoff, on the other hand, dynamics ought not to address a single problem, but rather tackle the most general class of dynamical systems."[37]

Birkhoff was part of a new generation of American mathematicians that emphasized pure mathematics over the applied forms that had historically dominated the field. Yet Birkhoff was a curious mixture of both the pure and applied schools; while he spent much time pursuing applications in celestial mechanics, he never once actually computed an orbit, according to Aubin.[38]

Birkhoff's 1927 text encapsulated more than a decade and a half of his work on general dynamical systems, the three-body problem, and the special case of two degrees of freedom to which he paid particular attention. However, it did not include what many considered to be his greatest work in this area—that pertaining to "ergodic" theory—which would not come for several years. When his proof on this subject was published in 1931, Birkhoff was justifiably proud, claiming that it embodied "a strong, precise result which, so far as I know, had never been hoped for."[39]

Others certainly shared that assessment. John Franks, for example, called his work on ergodics "a contender for the greatest theorem of the 20th century."[40] This work was "a remarkable tour de force," asserted Norbert Wiener.[41] "It is one of the greatest triumphs of recent mathematics in America or elsewhere that the correct formulation of the ergodic hypothesis, and the proof of the theorem on which it depends, have both been found by the elder Birkhoff at Harvard."[42]

Although Birkhoff may have "found" the proof, as Wiener put it, some of the ideas behind it dated back to the middle to late nineteenth century. At that time, the physicists James Clerk Maxwell and Ludwig Boltzmann were trying to formulate a kinetic theory of gases that described the behavior of a system consisting of a large number of particles (typically on the order of 10^{23}) in constant motion. Both Maxwell and Boltzmann believed that if the system is kept at a constant energy and allowed to evolve over an extended period of time, every possible configuration and state of motion will eventually be realized, meaning that, sooner or later, every point in "phase space"—an abstract space that encompasses all possible conditions or "microstates" of the system—will be visited. This assumption came to be known as the "ergodic hypothesis."

The hypothesis was later modified on various occasions to make it more precise. Mathematicians had determined, for instance, that a small set of points in phase space would never be visited and therefore had to be excluded from the outset. A further refinement was to acknowledge that you can never reproduce an exact configuration or return to an exact starting point. Instead, you draw a "neighborhood" around a given point, noting that you can come as arbitrarily close to that point as you want without getting there exactly. The amount of time spent in a given neighborhood, moreover, is proportional to its area or volume.

Though essential, these modifications did not irreparably change the basic concept that Boltzmann had incorporated into statistical mechanics some decades earlier, indeed making it a cornerstone of the new branch of physics that he founded in 1870. The mathematical soundness of this premise, however, was not firmly established until the early

1930s, when John von Neumann (who was then based at Princeton) and Birkhoff proved different versions of the ergodic theorem through dynamical arguments, though their approaches were otherwise quite dissimilar. Success for both von Neumann and Birkhoff hinged, in part, on finding a mathematically precise way of formulating the problem.

Birkhoff's interest in the subject was piqued, no doubt, by Poincaré's recurrence theorem, which was published in 1890. The theorem concerns systems that will, after the sufficient passage of time, return to—or come close to—their initial state. As an example, suppose that two chambers are connected but separated by an impermeable valve. The left-hand chamber is filled with gas, whereas the right-hand chamber is a vacuum. If the valve is then opened, gas from the filled chamber will diffuse into the previously empty chamber. Before long, the system equilibrates and both sides will have roughly equal concentrations of gas.

But it will not always remain thus, for Poincaré's theorem says that if we wait long enough, all of the gas will at some point return to the left-hand chamber, while the other chamber will be totally devoid of gas. This is a completely classical result, though it seems counterintuitive and wholly at odds with what we observe. The question is, how long do we have to wait? It turns out that the Poincaré recurrence time is so long that, from a practical standpoint, we never expect to see something like that happen. Yet, over the long run, that recurrence is inevitable all the same.

Birkhoff proved a more general version of the recurrence theorem en route to his ergodic theorem. (Both theorems, by the way, apply to systems of "finite measure," which basically means that the total areas in which the events occur are assumed to be finite.) A system that is ergodic will eventually pass through, or come close to, almost every initial condition, every point in phase space—save for the aforementioned excluded set—an infinite number of times. But the question of probability must also come into play. If you were a tiny gas particle, free to move about the chamber, not only would you visit every point infinitely often, but also the frequency of your visits would depend on a probability distribution. So the theorem tells you not only the places you will visit but also the odds of your being found at a particular spot at any given time.[43]

One of the main reasons people were interested in the ergodic theorem (and still are) stems from the fact that it makes a simple yet powerful statement, which goes well beyond the recurrence theorem: *The "time average" equals the "space average."* Before elaborating on those terms, let us note that systems for which this is true are now called ergodic. So Birkhoff's work, and von Neumann's as well, led to a new and simpler definition of the term.

To understand what the agreement between these two averages means, let us first explain what each of the averages refers to. We start with a simple, one-dimensional example—that of a circle with a random temperature distribution that fluctuates wildly from point to point. (The nuances of this idea are perhaps better illustrated with a two-dimensional example, such as the surface of a doughnut, but that situation is much harder to grasp.) In the case of the circle, the temperature is the only parameter we are considering—the only condition or state of concern here. (In more complicated systems, of course, there could be a large number of observables, all of which are required to characterize a particular state.) To get the space average, you would like to have observers positioned at infinitely close intervals, each making simultaneous temperature measurements all around the ring. We could easily derive an average temperature—or space average—by adding up all those numbers and dividing by the number of points. The mathematically precise way of doing this would be to define a function $f(x)$, which is equal to the temperature at a given point x, and then integrate that function over every point on the circle.

The space average is derived in a frozen or "static" situation, where measurements are simultaneous, made at a specific moment in time. Determining the time average is more complicated because time, in this case, is no longer frozen. Conditions can vary, and points on the circle are free to move around. To derive the time average, we start at an arbitrary point on the circle and measure its temperature. Because every point in this system is moving, by definition, we will measure its temperature at a new location on the circle—and at a new state—a tiny fraction of a second later, and we will repeat this process, over and over again, for a long time. If we are patient, and wait long enough, the number we get from averaging all those temperatures at various points on the circle will eventually reach, or converge on, a limit: a single number, which in this case happens to be the temperature obtained through the space average.

The reason these numbers turn out to be the same is that, owing to ergodicity—a feature of the system that eventually takes you all around the ring—the points you measure from the time evolution case end up being the same points you select for the space average measurement.

And the fact that the energy in this system stays constant ensures that the average temperature stays constant, too. In the end, you are making the same measurements from the same points. In one case—the space average—you are doing these measurements all at once. In the other case—the time average—it takes quite a while to visit all those spots. But either way, you would obtain the same constant value.

Establishing equality between the space and time averages—as both Birkhoff and von Neumann did—can be of great practical benefit. Depending on the problem at hand, sometimes it is easier to compute one of the averages than the other. But once you know one, you immediately know the other. In statistical mechanics, for instance, physicists often want to know the time average, which provides information about the trajectories of individual molecules, their positions over time, and their velocities. But if the system in question is a gas composed of, say, 10^{23} molecules, it is impossible to sample all initial conditions and solve Newton's laws of motion for every single molecule. So instead, you can apply the ergodic theorem and compute the space average, which can be worked out with pen and paper using a simple probability distribution. That is a very fruitful approach, which has been used, among other examples, to understand the dynamics of a Bose-Einstein condensate—a state of matter that can arise when dilute gases are cooled to extremely low temperatures.[44]

Birkhoff's theorem is sometimes called the pointwise ergodic theorem, meaning that you can pick almost any starting point—almost any initial condition in that abstract phase space—and over time you will get to almost every possible point, so that the average value you calculate will converge on the so-called space average. Note the use of the world "almost." Birkhoff proved that you could start from any initial condition or point except for those lying on a "thin set," which would be confined to a region of zero area in a two-dimensional space and a region of zero volume in a three-dimensional space. As time evolves, you will never get to any of these isolated spots either. Although this thin set (or "set of measure 0," as it is sometimes called) needs to be acknowledged for the sake of precision, its existence has no practical effect on the outcomes here.

For this approach to work—for the time average, starting from almost any initial condition, to equal the space average—a couple of conditions must be met. The system's overall energy, which correlates with temperature, must remain constant; it cannot vary in time. The system, as stated before, also has to be ergodic, which imposes constraints on how things move around in this phase or state space. Going back to the circle of our previous example, if you start at one point and move only in increments of 180 degrees around the circle, you will visit only two points, which is not nearly enough to get a good estimate. Satisfying the condition of ergodicity means that large chunks of space are not partitioned off and that—except for an essentially negligible thin set that is off limits—you will, after moving from point to point in a seemingly haphazard fashion, eventually pass through almost every possible way station.

Von Neumann's "mean ergodic theorem" is less detailed than Birkhoff's. It does not tell you anything about convergence from a particular starting point or about convergence along a particular trajectory; it addresses only convergence *in the mean*, demonstrating that the time average, as a general matter, converges toward the space average. Both theorems, however, represent major advances. Both were sufficient to justify the intuition of physicists like Maxwell and Boltzmann, thereby providing a rigorous mathematical foundation for many of the principles of statistical mechanics, although not in the exact form in which these ideas were originally advanced. And both theorems contributed to the founding of a new field in mathematics, "ergodic theory."

John Franks contends that owing to its greater specificity, the Birkhoff theorem has supported most of the subsequent work in mathematics.[45] "Both theorems and their generalizations are fundamentally important to the subject," says UCLA's Terence Tao, a winner of the Fields Medal, sometimes called the Nobel Prize of mathematics, which is awarded to mathematicians who are no more than forty years old.[46]

"While the mean ergodic theorem is, strictly speaking, weaker than the pointwise ergodic theorem," Tao adds, "it is easier to prove and, as a consequence, there have been many more generalizations of the former theorem than of the latter." For example, Tao's acclaimed work with University of Cambridge mathematician Ben Joseph Green—in which they solved a problem relating to progressions of evenly spaced prime numbers—"used a deep generalization of the mean ergodic theorem known as the Furstenberg recurrence theorem," Tao says.[47]

Despite the fact that both Birkhoff and von Neumann, through their respective ergodic theorems, made important contributions to mathematics and physics, with more than enough glory to go around, a controversy arose over the question of who proved the ergodic theorem first—a situation complicated by the order in which the two papers were published. Before the two mathematicians embarked on their proofs, Bernard Osgood Koopman, a former Birkhoff Ph.D. student then at Columbia (and relative of William Fogg Osgood), supplied a critical bit of mathematics that came to light in a paper he published in May 1931. After seeing Koopman's note in the *Proceedings of the National Academy of Sciences* (*PNAS*), von Neumann figured out a way to proceed and finished his mean ergodic proof by September. A month later, Birkhoff attended an event at Princeton, where von Neumann and Koopman told him of von Neumann's theorem, which incorporated Koopman's method, called "proof of the quasi-ergodic hypothesis." Von Neumann presented his proof in Cambridge in December 1931, "and in the discussion," von Neumann said, "Birkhoff informed us that he had another proof, which showed even somewhat more than mine: Instead of mean convergence, he could prove convergence everywhere excepted on a set of measure 0. (In a paper . . . , I show that the physical statement of the q.E.H. [quasi-ergodic hypothesis] requires mean convergence and not more.)"[48]

According to von Neumann, during dinner at the Harvard Club he asked Birkhoff when his paper would appear, as his was scheduled for the January 1932 issue of *PNAS*. (Von Neumann had taken extra time before submitting his paper, while he waited for Koopman and Marshall Stone, another former Birkhoff student who was then teaching at Harvard, to review his manuscript.) Birkhoff told him that his proof might come out in the December issue, though he was not sure. Birkhoff refused to withhold his paper until von Neumann's was published, von Neumann said, promising instead that "he will acknowledge my priority in due form."[49]

Birkhoff's ergodic proof was, in fact, published in the December 1931 issue of *PNAS*, whereas von Neumann's proof did not come out until January 1932. Birkhoff's acknowledgment to von Neumann, which appeared on the first page of his paper, merely said: "The important recent work of von Neumann (not yet published) shows only that there is convergence in the mean."[50]

Von Neumann was not satisfied with Birkhoff's "quotation of my result," noting that Koopman, Stone, and Lefschetz, among others, considered it "absolutely insufficient . . . The real reason they give is that it does not show to any person, uninformed about the real history of these things, who of Birkhoff and myself got the other started, that which one of us attacked the unsolved q.E.H., and which one found an independent new proof after he knew that it was solved."[51]

Birkhoff made amends in the March 1932 issue of the *PNAS* in a paper he wrote jointly with Koopman, which spelled out the history more explicitly, stating that "the first one actually to establish a general theorem bearing fundamentally on the Quasi-Ergodic Hypothesis was J. v. Neumann." The paper explained that Birkhoff had seen this result in October 1931, obtaining his own result "shortly thereafter... by entirely new methods."[52] This paper clarified the record in print, "apparently to the satisfaction of von Neumann," according to Joseph D. Zund, an emeritus mathematics professor at New Mexico State University. Von Neumann soon followed up with several more papers on ergodic theory, whereas Birkhoff, curiously, did no further research in this area. Birkhoff did, however, discuss applications of ergodic theory to physics, lamenting "that physicists working in statistical mechanics had not yet taken proper notice of the importance of ergodic theory in their work."[53]

Both mathematicians apparently moved on from the controversy, leaving that matter to rest. Birkhoff's son Garrett soon became a collaborator, as well as a friend, of von Neumann's. Meanwhile, estimations of the elder Birkhoff's mathematical prowess, in the wake of his ergodic proof, were never higher. Indeed, when Abraham Flexner, the director of the Institute for Advanced Study, which was founded in Princeton in 1930, began looking for someone to head the institute's new School of Mathematics, he put Birkhoff at the top of the list and started recruiting him vigorously. Birkhoff, for his part, came very close to taking that position. In fact, he accepted an offer from Flexner twice before retracting his acceptance on both occasions, finally deciding in 1932 to remain at Harvard.

Although his reasons for staying were not entirely clear, Birkhoff was given an honorary degree at the 1933 commencement, when he was also named the Perkins Professor of Mathematics. The math historian Steve Batterson speculates that Birkhoff might have been "reluctant to join an enterprise funded and directed by Jews," an issue explored later in this chapter. (Flexner was Jewish, as were the founders of the Institute for Advanced Study: New Jersey businessman Louis Bamberger and his sister Caroline Bamberger Fuld.) Another reason for staying might have been his son Garrett, a promising mathematician who was soon to begin a three-year research fellowship at Harvard. "The father may have foreseen parts of his son's trajectory and wished to be at Harvard as it unfolded," writes Batterson.[54]

In 1933, Birkhoff went off in an entirely new direction—one that a number of his colleagues did not fully embrace. It had nothing to do with dynamical systems or ergodic theory but instead involved "aesthetic theory." In a series of lectures and in a book titled *Aesthetic Measure*, Birkhoff tried to find an objective, quantitative way of evaluating an artwork's attractiveness. His theory can be boiled down to a simple equation: $M = O/C$, where M is the aesthetic measure of an art or music piece, O is the piece's "order" (which relates to how harmoniously its parts fit together), and C is its complexity.

His interest in analyzing art and music had been longstanding, according to Morse, and Birkhoff gave a hint as to where that interest sprang when a musician asked him about the point of studying mathematics. "One should study mathematics," Birkhoff replied, "because it is only through mathematics that Nature can be conceived in harmonious form."[55] In addition to writing a book on the subject of aesthetics, Birkhoff also took a half-year leave from Harvard to gather material in Europe and Asia.

Birkhoff was by no means the first mathematician to veer off in a new research direction that seemingly had little connection to his prior work. "It was quite common in those days for scientists to theorize on other subjects," Garrett remarked. "Simon Newcomb, I think, wrote 500 papers on mathematics and almost as many on other subjects. It was common for scientists to speculate about the nature of knowledge."[56] Going back farther still, Pythagoras developed a scientific theory of music (being perhaps the first person to do so), showing that in pleasing melodies there are simple numerical ratios between the notes. Birkhoff was trying to do something similar, although broader, in devising a mathematical theory for the fine arts in general.

As described in a 1934 magazine article, Birkhoff applied his aesthetic criteria to polygons (with the highest rating given to a square), paintings, and poems.[57] He presented a formula for appraising poetry that expressed the "order" in terms of alliteration and assonance, rhyme, musical sounds, alliterative excess, and an excess of consonant sounds. Based on these criteria, Alfred Lord Tennyson's poem "Come into the Garden, Maud" was among the top-rated poems on Birkhoff's scale, scoring a 0.77. Birkhoff applied that same formula to a poem he wrote, which garnered a 0.62:

> Wind and wind the wisps of fire,
> Bits of knowledge, heart's desire;
> Soon within the central ball
> Fiery vision will enthrall.
>
> Wind too long or strip the sphere,
> See the vision disappear![58]

Not everyone was impressed. Joseph Doob, for instance, sat in on the course "Aesthetic Measurement" when he was a Harvard mathematics graduate student (earning his Ph.D. under the direction of Birkhoff's former student Joseph Walsh). Doob constantly questioned Birkhoff about his formula for assigning numerical values to works of art, asking whether he actually liked the art pieces that earned high numbers on his scale. "In my youthful brashness I kept challenging him on the absence of definitions, and he finally came to class one day, carefully focused his eyes on the ceiling and said that those not registered in the class really had no right to attend. I took the hint and attended no further classes." Doob was pleased to report that the professor bore him no grudge, asserting that it was through Birkhoff's influence that he later received a two-year National Research Council fellowship.[59]

The British mathematician G. H. Hardy was also no fan of Birkhoff's latest research foray. When Garrett Birkhoff came to Cambridge, England, in 1932, Hardy asked him, "How is your father coming along with his aesthetic measure?" Garrett told him that the book had been completed. "Good," Hardy replied. "Now your father can get back to real mathematics."[60]

But Birkhoff soon took on more administrative duties, becoming dean of the Harvard faculty in 1935, which naturally afforded him less time for research, though he continued to publish papers throughout the decade on dynamical systems, the four-color theorem, and other topics. In 1943, Birkhoff published a paper that presented a novel theory of gravitation. His model assumed a flat four-dimensional spacetime, unlike the curved spacetime in Einstein's description of gravity. The gravitational potential in Birkhoff's model was governed by linear differential equations, as opposed to the nonlinear differential equations that abounded in general relativity theory. With the benefit of hindsight, it now appears that Birkhoff's attempt was misguided; Einstein's formulation agrees with the observations better in almost every respect.

Despite Birkhoff's prodigious contributions to math, some researchers found little value in his attempt to rewrite gravitational theory. Morse, however, cast this effort in a somewhat more charitable light. "It has been stated ... that this theory does not provide for the identity of gravitational and inertial mass in as effortless a manner as the theory of Einstein, and this seems to be a defect from which linear theories must suffer," he commented. "[But] Birkhoff inherited from Poincaré the sentiment that no single mathematical theory of any phenomenon deserves the exclusive attention of physicists."[61] On this matter, Birkhoff might also have subscribed to the dictum, put forth by the mathematician David Hilbert, that physics was too difficult to be left to physicists alone.

Birkhoff continued to work on his alternative theory of gravitation, though he did not win many adherents, and his progress in this area was ultimately stalled by failing health. He took a nap on Sunday, November 11, 1944, before he and his wife were supposed to go to Garrett's house. Although he may not have been feeling tired, he had been advised by his doctor to take it easy since an incident in the spring, six months earlier, when he collapsed while walking up a hill, en route to the house of David Widder, a former student who was then a Harvard math professor. Birkhoff never woke up from that nap.

In his passing, many remembered his remarkable and varied accomplishments, as well as those things he had never quite finished. The mathematician E. T. Bell recalled that "Birkhoff said shortly before his death that in spite of all his efforts, one of which I witnessed in 1911, to crack the four-color problem wide open, he had not even scratched it." Birkhoff's friend Vandiver did not consider that "such a situation is unusual with mathematicians in their later days. Many individuals who have spent most of their lives on mathematical research will often regret deeply that certain problems on which they had spent a great deal of time were never solved by them."[62]

Had he survived for another few decades, Birkhoff might have felt vindicated by a proof of the four-color theorem based on the reducibility approach that he had pioneered. Kenneth Appel and Wolfgang Haken made use of that strategy, as well as a computer. In 1976, after four years of computations and 1,200 hours of computer time, Appel and Haken had successfully carried out Birkhoff's program, proving the four-color theorem in the process. This was the first time a major mathematical theorem was verified by computer, and, as such, the proof was not without controversy. But in the decades that followed, confidence in that result has grown, and most mathematicians now believe that the four-color theorem was indeed proved, about sixty years after Birkhoff paved the way.

At the time of his death, however, no one held it against Birkhoff that he had never fully solved the problem himself. Instead, the accolades poured in from all quarters: "Thus passed America's leading mathematician, who had been recognized here and abroad more widely than any other mathematician America has yet produced," his fellow Harvard mathematician Edwin B. Wilson wrote in *Science* in 1945.[63]

Rudolph Langer, a former Birkhoff student who later joined the University of Wisconsin faculty, wrote: "Where he set his hand to the plow, the furrow never failed to be deepened, if, indeed, its direction was not soon materially altered. His genius guided him in the divining of many ways to significant and previously inaccessible results—ways in which others have been content to follow him."[64]

"During the major part of his life Birkhoff was the acknowledged leader of American mathematics," said Morse, pointing to Birkhoff's nomination as president of the International Congress of Mathematicians in 1940 (which was itself postponed due to the war) as evidence of his lofty stature.[65] His devotion to Harvard mathematics and to American mathematics as a whole was absolute, unwavering, and probably excessive. This almost "religious devotion to American mathematics as a 'cause'" was, according to Veblen, characteristic of many of Birkhoff's predecessors (such as Benjamin Peirce), as well as many contemporaries. "It undoubtedly helped the growth of science during this period," Veblen added.[66] But many feel that Birkhoff went too far in this direction, exhibiting extreme nationalism at times, as well as anti-Semitism.

The latter issue raised by Veblen (a close friend of Birkhoff's), as well as by many others, is an important one, warranting some careful consideration. In a speech at the 1938 celebration of the American Mathematical Society's fiftieth anniversary, Birkhoff revealed some of his attitudes regarding the wave of refugees arriving on America's shores. On the one hand, Birkhoff paid tribute to eminent European mathematicians of "high ability" who had come to the United States "largely on account of various adverse conditions." His list included Emil Artin, Richard Courant, John von Neumann, Hermann Weyl, and Oscar Zariski. But Birkhoff did not stop there. "With this eminent group among us," he continued, there inevitably arises a sense of increased duty toward our own promising younger American mathematicians. In fact most of the newcomers hold research positions, sometimes with modest stipend, but nevertheless with ample opportunity for their own investigations and not burdened with the usual heavy round of teaching duties. In this way the number of similar positions available for young American mathematicians is certain to be lessened, with the attendant probability that some of them will be forced to become "hewers of wood and drawers of water." I believe we have reached a point of saturation where we must definitely avoid this danger.[67]

Birkhoff's speech was controversial, because many people believed that when he spoke of "newcomers" he was really referring to Jews. He warned that if foreign mathematicians took all the plum research positions, younger Americans would be relegated to teaching introductory courses and general grunt work, which is presumably what he meant by hewing and drawing. "In Birkhoff's defense, it should be said that the situation of gifted young American Ph.D.'s in mathematics was desperate in 1938," explained the mathematics writer Constance Reid. Some of these researchers had to support families on stipends of $1,000 per year or less.[68]

Saunders Mac Lane, a mathematician at Harvard and the University of Chicago, benefited early in his career from Birkhoff's policy of giving young American mathematicians precedence over refugees. Mac Lane was offered (and accepted) a two-year Benjamin Peirce Instructorship in 1938 that might just as well have been offered, he said, to "any one of a half dozen well qualified émigrés." Mac Lane remained on the Harvard faculty through 1947 and in that time did not recall Birkhoff making any specific statements about Jews. Yet he claimed to have "dependable evidence that [Birkhoff] thought that Jewish mathematicians stopped doing research early . . . It seems likely," he says, "that Birkhoff shared the somewhat diffuse and varied versions of anti-Semitism held by many (most?) of his contemporaries." These attitudes, Mac Lane correctly pointed out, "were not limited to mathematics or to universities, but were present in the larger society."[69]

One sad fact of that era was that Harvard, Yale, Princeton, and other schools had institutionalized discrimination against Jews, limiting the number of Jewish students and faculty. Although Birkhoff did not originate this practice, he clearly embraced it. When asked by Oscar Zariski in 1926 whether it was difficult for a Jew to become a Harvard student, Birkhoff replied: "No, not at all, although naturally we keep a certain proportion. The Jewish population is about 3%, 4%, 5%, so naturally we admit only 3%, 4%, 5%."[70]

It is often mentioned that Birkhoff aggressively went to bat for a Harvard math Ph.D., a complex analyst named Wladimir Seidel, urging Rochester's mathematics chair to hire him. When the chair balked, Birkhoff countered: "I know you hesitate to appoint the man I recommended because he is a Jew. Who do you think you are, Harvard? Appoint Seidel or you will never get a Harvard Ph.D. on your faculty."[71]

One might point out that Birkhoff did not offer that same Ph.D. a tenure-track position at Harvard. Birkhoff did, however, approve the appointment of some Jewish mathematicians for fellowships and temporary positions, including Stanislaw Ulam, who came to Harvard at Birkhoff's suggestion as a member of the Society of Fellows from 1936 to 1939, spending an additional year at the university as a lecturer. The talented Ulam (who is discussed at greater length in Chapter 6) was a protégé of Birkhoff's, and the distinguished professor took an interest in the young mathematician's career. When Birkhoff heard of a math opening at Columbia, he urged Ulam to apply, saying that he would recommend him highly. Ulam replied that he was happy at Harvard, but Birkhoff said, "You don't know how the game is played here. When an opening like this comes along, you have to apply." In his letter of recommendation to Columbia's math chair, Birkhoff described Ulam as a "brilliant and original mathematician." Birkhoff also told the Columbia scholar, in the same letter, that he had read his latest paper, "and your result is trivial. Here is a three-line proof of it." Ulam, not surprisingly, did not get the job and instead took a position at the University of Wisconsin, before later moving to Los Alamos.[72]

Although Birkhoff went out of his way to help some Jewish scholars like Ulam, none was hired for a permanent math faculty position during his three-plus decades at Harvard (from 1912 to 1944). Indeed, when Harvard hired Zariski in 1947, three years after Birkhoff's death, he became the first Jew ever to receive tenure in the department. (In all fairness, it should be said that Birkhoff had wanted to hire Zariski earlier but was unable to do so due to a hiring freeze during the war.)

Norbert Wiener, a Harvard math Ph.D., called Birkhoff "intolerant of possible rivals and even more intolerant of Jewish rivals. He considered that the supposed early maturity of the Jews gave them an unfair advantage at the stage at which young mathematicians were looking for jobs, and he further considered that this advantage was particularly unfair, as he believed that the Jews lacked staying power."[73] (Elsewhere, Birkhoff had characterized Jewish mathematicians as "early bloomers," who peaked early and stopped producing once they got tenure.)[74]

At first, Wiener considered himself too unimportant to attract Birkhoff's attention, "but later on, as I developed more strength and achievement, I became his special antipathy, both as a Jew and, ultimately, as a possible rival." Wiener concedes that "he was not a very amiable young man" and was indeed "an aggressive youngster," which might also have been strikes against him.[75]

In this case, it is not obvious that Wiener—soon to become recognized as one of the most capable mathematicians in the country—did not get an offer from Harvard simply because he was Jewish. Other factors, including the fact that he rubbed many people the wrong way, may have influenced the hiring decisions as well. But it is undeniable that in 1934 Birkhoff opposed Solomon Lefschetz's election as president of the American Mathematical Society at least in part because he was Jewish. In a letter to his friend Roland Richardson, who was then secretary of the society, Birkhoff wrote: "I have a feeling that Lefschetz will be likely to be less pleasant even than he had been, in that from now on he will try to work strongly and positively for his own race. They are exceedingly confident of their own power and influence in the good old USA . . . He will be very cocky, very racial, and use the *Annals* [*of Mathematics*, which Lefschetz edited from 1928 to 1958] as a good deal of racial perquisite. The racial interests will get deeper as Einstein's and all of them do."[76]

Although Birkhoff was not without company, statements of this sort would suggest that he had a far more jaundiced view of his Jewish colleagues than most of his contemporaries. Albert Einstein, to whom Birkhoff alluded above, called Birkhoff "one of the world's great anti-Semites."[77] Mac Lane dismissed that quote as "worthless," since "Einstein did not then carefully follow the American academic scene," and "Birkhoff had a (then well-known) competing theory of relativity."[78] Nevertheless, Mac Lane's contention that Birkhoff's views merely reflected society at large might have been overly charitable. When Harvard president James Bryant Conant sent a representative from Harvard in 1936 to celebrate the 550th anniversary of the University of Heidelberg—which was then a center for "Aryan physics" and almost entirely under Nazi control—he might have chosen any of a number of people or declined the invitation outright. But he picked Birkhoff, who was then the faculty dean, to represent Harvard in Heidelberg. "At the ceremonies, [Birkhoff] was in the company of Nazi propaganda minister Josef Goebbels ... and SS chief Heinrich Himmler," writes the historian Stephen Norwood.[79]

While that unfortunate association may not necessarily shed light on Birkhoff's personal beliefs, there is still ample evidence to suggest that his anti-Semitic tendencies went beyond the norms of the era and far beyond the norms in American mathematics at that time. And if that were the case, in assessing Birkhoff the man, we ought to consider the whole package, weighing his brilliant mathematical achievements against some aspects of his personality that may have been less commendable.

As Richard Courant, a German Jewish émigré who founded the Courant Institute of Mathematical Sciences at New York University, said of Birkhoff: "I don't think he was any more anti-Semitic than good society in Cambridge, Massachusetts, used to be. His attitude was very common in America at that time. I think Birkhoff was narrow and certainly he was wrong—but he was certainly a very good mathematician."[80] Yale mathematician George Daniel Mostow, a former president of the American Mathematical Society, as well as a former Harvard math Ph.D. who took courses with Birkhoff, put it more bluntly: "Brilliance and big brains does not mean a big heart."[81]

It is gratifying to know that Birkhoff's worst fears about the dangers to American mathematicians posed by a rising tide of refugees were never realized. The domestic mathematical community was able to absorb "a substantial number of European mathematicians without serious indigestion," Veblen noted, showing that U.S. mathematics was strong enough that it could be less nationalistic.[82]

The presumed "conflict of interest between young American mathematicians and European refugees never led to much jealousy or hostility that I could see, because a sense of fair play and mutual sympathy dominated most decisions," wrote Garrett Birkhoff, who appeared to have a more modern and compassionate perspective than his father. This sentiment was coupled with the conviction that the United States offered the best place to do research in science and math, the younger Birkhoff added. "So who could complain?"[83]

Indeed, large numbers of Jewish scholars came to the United States from abroad, and far from weakening American mathematics, they clearly and incontrovertibly strengthened it. Exact numbers are hard to come by, though one estimate held that about 150 European mathematicians migrated to escape the Nazis, the majority of whom were Jewish.[84]

Some of these people were already established figures in their fields, while others were just starting their careers. Yet the combined impact of their presence was large and positive, argues University of New Mexico mathematician Reuben Hersh. "By driving the Jewish mathematicians and physicists from Europe to America, Hitler gave the U.S. a present more valuable than anything you can think of."[85]

"The young men (they were almost all men), about whose future Birkhoff worried, did not become 'hewers of wood and drawers of water,'" wrote Lipman Bers, a Latvian-born Jew (once advised to change his first name to "Lesley") who immigrated to the United States in 1940 and later taught at Syracuse, the Institute for Advanced Study, New York University, and Columbia. "On the contrary," says Bers (whose work is discussed in Chapter 5), "they became the leaders of American mathematics, and under their leadership all traces of xenophobia and anti-Semitism disappeared from mathematical life."[86]

As an indirect legacy of Birkhoff, ironically, it seems that the young corps of U.S.-born mathematicians that he was so determined to protect proved worthy of his efforts. Through their collective wisdom and tolerance, these same people, representing the next generation in American mathematics, helped elevate their field, while improving their country at the same time.

4

ANALYSIS AND ALGEBRA MEET TOPOLOGY: MARSTON MORSE, HASSLER WHITNEY, AND SAUNDERS MAC LANE

In contemplating the legacy of George David Birkhoff, who has been called (justifiably so) a "towering figure" in American mathematics,[1] we ought to consider more than his intellectual achievements or the personal shortcomings described in Chapter 3. We should also take into account his leadership role in the field, as well as the remarkable group of graduate students he trained at Harvard. In the preface of this book, we used the metaphor of a river to describe the flow of ideas in mathematics. But one can also invoke the notion of a family tree, starting with luminaries like Leonhard Euler and Carl Friedrich Gauss, seeing which mathematicians they trained personally and those whom their students trained, and charting how things branched out from there. Such an undertaking, called the Mathematics Genealogy Project, is currently under way. Based at North Dakota State University, in association with the American Mathematical Society, the project shows Birkhoff to have had forty-six Ph.D. students (all but a few at Harvard and the affiliated Radcliffe College) and more than 7,100 "descendants," a growing number that consists of his students, his students' students, and so forth.

In Birkhoff's case, it is not just the *number* of students he trained but the *caliber* of those students and the important branches of mathematics that have sprouted and grown through their collective efforts, including topology—the study of shape in the most general (and often abstract) sense—which is the principal topic of this chapter. Four of Birkhoff's students—Marshall Stone, Joseph Walsh, Charles B. Morrey, Jr., and Marston Morse—went on to become presidents of the American Mathematical Society. Three former students—Morse, Hassler Whitney, and Stone—won the National Medal of Science for mathematics, and they and other students racked up numerous awards in the course of their distinguished careers.

Stone, for example, started out in classical analysis, pursuing the theory of differential equations, in keeping with the interests of his research supervisor, Birkhoff. But Stone quickly switched to more abstract realms of algebra and topology. He taught at Harvard from 1927 through 1931 and then again from 1933 to 1946, taking time off to serve his nation in World War II. In 1930, Stone and John von Neumann published their famous uniqueness theorem, which provided a key mathematical underpinning to the quantum theory of physics. After the war, Stone moved to the University of Chicago. In addition to his notable accomplishments in pure mathematics, he was also a legendary administrator, famous for building up Chicago's math department in the 1940s and 1950s—and making it one of the nation's finest at the time—during a tenure subsequently referred to as the "Stone Age."

Following in the footsteps of Birkhoff, Morrey made his mark in analysis, pioneering new techniques for solving linear and nonlinear differential equations. The German-American mathematician Kurt Friedrichs first met the young Morrey at a New York meeting of the American Mathematical Society, when "this inconspicuously-looking boy came up to me and said modestly that . . . he had been working on partial differential equations and had solved such and such a problem." Friedrichs was dumbfounded when he realized the problem was one that "many of us had worked on for years and years. I couldn't believe it. Oh yes, Morrey [was] powerful."[2] Indeed, the tools introduced by Morrey provided access to problems that were previously beyond reach, opening up new doors in mathematics. He spent forty years, practically his entire career, on the faculty of the University of California, Berkeley.

Walsh, by contrast, was a Harvard man through and through. He got a bachelor's degree and a Ph.D. from the university and remained on the faculty from 1921 until his retirement in 1966, save for the four years he spent in the U.S. Navy during World War II. He, too, followed in the tradition of Birkhoff and William Fogg Osgood, focusing on real and complex analysis. Walsh published nearly three hundred research papers in his long career and supervised thirty-one doctoral students at Harvard—Lynn Loomis (who later joined the Harvard faculty) and Joseph Doob notable among them. A celebrated, as well as tireless, teacher, Walsh was devoted to his students and known for his meticulously prepared lectures on a broad range of subjects. When asked for pedagogical advice by a former student just beginning his teaching career, Walsh offered the following suggestion: always start writing from the upper left corner of the blackboard.[3]

As for Morse, he is one of the main subjects of this chapter, along with Whitney, another former Birkhoff student, and Saunders Mac Lane, who did not study under Birkhoff but benefitted from his hiring philosophy and learned from him as a junior faculty member. These three people radically changed the practice of topology, with the latter two bringing the field closer to algebra, thereby facilitating a wave of advances.

Marston Morse

Morse got his Ph.D. from Harvard in 1917, choosing for his dissertation topic a problem that combined analysis and topology—two fields that he sought to connect throughout his career. He was still at it, in fact, nearly fifty years later, noting that his "mathematical objective is to relate topology, globally, to local analysis and geometry. This research will go on," he affirmed.[4]

Morse left Harvard in 1917 to fight in World War I, returning to the university in 1919 as a Benjamin Peirce Instructor. He then taught for a few years at Cornell and Brown, rejoining the Harvard faculty in 1926. A year before coming to Harvard, in his 1925 paper "Relations between the Critical Points of a Real Function of n Independent Variables," Morse laid out the first installment of a new mathematical theory that would become his life's work.[5]

He elaborated on this theme, laying the foundations for what is now called Morse theory, in a book he wrote during his Harvard years, *The Calculus of Variations in the Large*.[6] The calculus of variations is a field of mathematics that, simply put, involves finding the "points of equilibrium" of functions, including the minima and maxima. For example, in a space of curves, such as all the curves that can be drawn on the surface of a sphere, mathematicians often look at curves where the length function achieves a minimum, which brings up the notion of "geodesics." Although the shortest distance between two points on a plane is a line that is readily identified, the problem is less obvious on complicated surfaces where many solutions may exist. In a space of surfaces, similarly, mathematicians can use the area function to identify minimal surfaces. The term "in the large" refers to global properties of a surface; the term "in the small," by contrast, refers to local properties in the neighborhood of a point on a surface. "Variations" has a rather intuitive meaning in this context, as it involves taking a curve (or surface) and gradually changing its length (or area) until you identify the curves of maximum or minimum length (or the surfaces of maximum or minimal area).

Morse theory offered a bold synthesis of analysis and topology. At the time, many mathematicians viewed topology primarily as a way of solving problems in analysis—in other words, a way of solving differential equations. For example, a "closed geodesic," such as a great circle (like the equator) on a sphere, is a solution to a second-order differential equation. Finding closed geodesics, therefore, is akin to solving differential equations. While acknowledging the value of that strategy, Morse also turned the approach upside-down by showing that his theory, which was based on analysis, could solve problems in topology as well.

Writing in 1977, the Fields Medal–winning mathematician Stephen Smale rated Morse theory as "the single greatest contribution of American mathematics (perhaps excluding more recent contributions for which time has been too short to assess sufficiently)."[7] Raoul Bott, who supervised Smale's thesis at the University of Michigan before coming to Harvard, wrote in 1980 that Morse theory was by then so established—so "natural and inevitable"—that it was hard to imagine "what a tour de force it was in the 1920s" when Morse first came up with it.[8]

Math historian Joseph D. Zund was similarly effusive in his assessment of Morse theory: "While it is not unusual for mathematicians to propose new theories based on an abstraction or a refinement of previously known results, very few have created a new theory where none existed before. Morse theory was such a theory, and in creating it Morse followed in the heroic traditions of Henri Poincaré and Morse's mentor George D. Birkhoff."[9]

Morse theory, put in the simplest possible terms, offered a new way to classify topological objects or "spaces," especially for higher-dimensional spaces that are not easily characterized. Topology itself is an extremely general though still very powerful way of describing shapes. In ordinary Euclidean geometry—the kind of geometry people are typically introduced to in high school—parallel lines never intersect, and the angles of a triangle always add up to 180 degrees. Two triangles are considered equivalent, moreover, if the sides are of the same length and all the angles are equal, so that one triangle exactly coincides with the other. "On the other hand," Morse explained, "in topology each triangle is regarded as equivalent to each other triangle. One disregards the notion of rigidity, so basic in Euclidean geometry, and notes that any given triangle can be moved and distorted so as to coincide with each other triangle."[10] This equivalence is by no means restricted to triangles. A triangle is topologically equivalent to a square, hexagon, and circle, just as a tetrahedron (a two-dimensional surface made up of four triangles) is topologically equivalent to a cube and sphere. Two shapes are considered equivalent in this sense if one can be transformed into the other by bending, stretching, or shrinking, but not by cutting.

For the same reason, a doughnut (or "torus") is topologically distinct from a sphere because it has a hole and a sphere does not. You cannot start with a sphere and end up with a doughnut without cutting a hole.

A doughnut, it turns out—especially one standing vertically (like an upright inner tube)—provides a relatively simple picture of what Morse theory is all about. Morse called his idea "critical point theory" because it focuses on the critical points of a function, linking those points to the overall topology of a "manifold"—a kind of topological space, central to both geometry and physics, that is described more fully in Chapter 6. The aforementioned vertical doughnut is a simple example of a manifold defined by a "height function" that assigns a number (corresponding to the height) to every point on the horizontal plane. A doughnut of this sort has four critical points, which, loosely speaking, are located at places where the shape of an object—or of a more general surface or space—abruptly changes in the course of moving from one end to the other. At these transitional spots, the surface is flat, meaning that planes tangent to the surface would be parallel to the tabletop on which the doughnut stands.

The first critical point, starting at the top, sits at the very peak of the doughnut—the maximum, in math parlance. The second critical point lies at the top of the doughnut's inner ring at a so-called saddle point, which we will define shortly. The third critical point lies at the bottom of the doughnut's inner ring at another saddle point on this surface. The fourth critical point lies at the doughnut's base: a single point, or minimum, that rests squarely upon our imagined tabletop.

What is important here is not just the number of critical points but also their nature. Critical points can be categorized by the index, which indicates the number of independent directions down. For that reason, Morse called it the "index of instability," because it tells you how many distinct ways a ball could descend if you placed it at a critical point and then gave it a slight nudge.[11]

Returning to our example of an upright doughnut, at the uppermost critical point (the maximum), the index is 2 because there are two independent (and perpendicular) directions down: one going down the middle of the outer ring, the other moving toward the hole in the center. The second critical point, the upper saddle, has an index of 1 because from that spot you can go down the middle of the ring or move in a perpendicular direction up toward the top. One direction goes down, and the other up, which is why the index is 1. The third critical point, the lower saddle, is similar, only the situation is reversed: moving along the inner ring will bring you up, while moving in the perpendicular direction takes you down. With just one way down, the index here is again 1. Finally, the fourth critical point at the base has an index of 0; no direction will take you down because you are already sitting at the bottom. Instead, two directions take you up.

It is also a fact that the number of independent directions up and down at each critical point adds up to the dimensionality of the surface, which is 2 in the doughnut example. The indices tell us the directions you can move up or down at key junctures on a surface, which provides an indication of how the critical points are connected and therefore tells us how the entire surface is put together. You can thus describe this surface simply by spelling out the sequence of its indices at the critical points: 2, 1, 1, 0. That is all the information you need from a topological standpoint to uniquely define this object as a doughnut (or two-dimensional torus). Using Morse theory, this same approach applies to other smooth surfaces—and even not-so-smooth surfaces—in any dimension you pick. As you move across the surface, you just have to keep track of all the critical points, as well as the index of each critical point. From that sequence you can determine its topology—be it a torus, sphere, or some other shape.

Everything else you see while wandering around the surface is pretty much irrelevant. You could think of this approach as a movie in which absolutely nothing of interest happens, except at a few key moments. These infinitesimally brief moments would take place exactly when you pass the critical points.[12]

Morse theory would be of limited value if it applied to just simple surfaces like doughnuts, which we could draw or build and then pick out the critical points by sight. If this methodology worked only for such surfaces, it still would have taken a brilliant insight on the part of Morse to realize that an object's topology can be divined from its critical points. But his approach is far more powerful than that because it also works for complicated surfaces and spaces of arbitrary dimension that are far too difficult for us to draw, let alone inspect by eye and thereby pick out the critical points. In situations like these, rather than relying on pictures, mathematicians instead tend to think in terms of functions.

We will choose an example here of something that one could readily draw to make it easier to visualize what is going on, even though the point in working with functions is that you are not limited just to the things you can draw. So we will take a function, $f(x,y)$, which, for any value of x and y on a coordinate plane, gives you a number z that corresponds to the height or elevation.

As Morse has suggested in his own lectures, you might think about this in terms of an island. There is an imaginary x-y plane sitting at sea level, and for every point on this x-y plane within the contours of the island, our function f will give us a point z that tells us the height of the island at that spot. Putting all those z's together—all the elevation points—as you move around on the x-y plane generates the actual surface of the island. This surface is bound to have some critical points: peaks (maxima), pits (minima), and some passes (or saddle points) located between the peaks or hills. One way to find the location of these critical points is to take the derivative of our function, which will be zero at the peaks, pits, and passes.

The derivative vanishes at places that are level—places where tangent planes to the surface are horizontal. What this means is that if you go to a critical point—be it a peak, pit, or pass—and move a tiny bit to the side in any direction, your height will not appreciably change. This is not the case on the side of our upright doughnut, where a small horizontal movement would produce a more significant vertical change. If you take the first derivative of the height function, it will be exactly zero at the critical points. For reasons that are harder to explain, and which we will not go into here, the second derivative of the height function tells us the index of the various critical points.

For more complicated, higher-dimensional "islands," this same approach of differentiation can identify the critical points. There can, of course, be more variables than just x and y, and as the number of variables increases, so too does the number of kinds of critical points. For a function of n (real) variables, there are "$n + 1$ possible types of critical points," according to Morse.[13] The index of a critical point can assume any value from 0 to n.

As a result of his theory, Morse brought the finely honed machinery of analysis and differential calculus into the newer, less mature field of topology. And his hopes of using analysis to solve problems in topology were quickly borne out. Indeed, an advanced form of analysis called geometric analysis was recently used to solve perhaps the most famous problem in topology, the Poincaré conjecture, a century after Henri Poincaré had sprung it upon an unsuspecting world.

To help us think mathematically about his imaginary island, Morse introduced three numbers (nonnegative integers): M_2, representing the number of peaks; M_1, representing the number of passes (saddles); and M_0, representing the number of pits. You will note that the subscripts in this case, 0, 1, and 2, correspond to the indices. So M_0, M_1, and M_2 are also the number of critical points with an index of 0, 1, and 2, respectively.

These numbers cannot assume any possible value but must instead satisfy certain relations—the so-called Morse relations, which include the famous Morse equalities and inequalities, all of which Morse proved to be true. An early example of such a relation comes from Birkhoff's minimax principle discussed in Chapter 3, which pertains to the existence of minima and maxima in length functions for curves covering a surface. The Morse relations go much further, showing how the numbers of maximum, minimum, and saddle points are related to each other. The relations also codified new insights on topological spaces—and rules for constructing such spaces—that did not exist before. For example, if we insist that the number of peaks, M_2, must be one or greater, then it holds, as Morse stated, that "the number of peaks and pits minus the number of passes equals one." Putting it more formally:

$$M_2 + M_0 - M_1 = 1, \quad M_2 \geq 1.$$

This means that an island could have one peak, one pass, and one pit, and it could have three peaks, four passes, and two pits, but no island can be constructed that has two peaks, two passes, and two pits.

If one were to consider not just an island but the entire Earth (smoothed out, so that the number of critical points is finite, and stripped of all water), Morse said, "we can prove that . . . the number of pits, plus the number of peaks, minus the number of passes equals 2" or, putting it in mathematical parlance: $M_0 + M_2 - M_1 = 2$. This relation, Morse added, can be extended to "surfaces of arbitrary genus"—or numbers of holes—"and to spaces of higher dimension."[14] In addition to these equalities, Morse also established a range of inequalities that contributed vital tools to topology. One of these inequalities was drawn from Birkhoff's minimax principle, but Morse found a way to generalize that to higher dimensions and to a wide range of spaces.

Following in Morse's footsteps, practitioners like Bott and Smale used Morse theory to solve problems in topology, analyzing the distribution of critical points to understand, and constrain, the topology of a manifold. Bott, for example, relied on Morse theory to prove his famous periodicity theorem, discussed in Chapter 7. Smale used Morse theory to prove the higher-dimensional Poincaré conjecture (valid for five or more dimensions). In particular, Smale constructed a function with only two critical points, proving that the manifold corresponding to this function must be a sphere.

Many other mathematicians have taken up this work, for many decades now, and it is still going strong. Morse theory now stands as a fundamental and indispensable approach in topology. "The theory is amazingly natural, so much so that it's automatically in the head of anyone who does anything in topology today," notes Harvard mathematician Barry Mazur.[15]

Morse theory has also led to extensive applications in modern physics. In 1983, for example, the physicist Edward Witten came up with an analytic interpretation and new proof of the Morse inequalities, which were especially adapted to quantum field theory.

It is probably safe to say that Morse did not know, at first, what his work would lead to, but he was guided from the outset by a firm conviction that analysis could be applied to questions in topology—the answers, in this case, proving to be nothing short of profound. Following in the tradition of Poincaré (of whom Morse was called a "mathematical grandson" in an honorary degree from the Sorbonne) and of his adviser Birkhoff (who might have himself been called Poincaré's mathematical son), Morse nevertheless set himself apart from his predecessors by veering off onto his own original path.

"Every real mathematician," Morse wrote, "senses some lack in the total mathematical picture," a gap or "great need which others have not sensed. Otherwise they would have done something about it. He has to prove that the need can be satisfied, in part at least." A starting point for Morse was a 1912 paper in which Birkhoff identified problems that neither he nor Poincaré could solve. "I took off from them," Morse said. "I sensed another entirely different lack in mathematics." He spent many fruitful decades successfully converting that "lack" into a dynamic field of mathematical inquiry.[16]

Morse was hard at this task in 1935 when he changed venues, leaving Harvard to join the mathematics faculty at the newly founded Institute for Advanced Study, where he worked alongside Albert Einstein, John von Neumann, Oswald Veblen, and Hermann Weyl. Morse officially retired from the institute in 1962, at the mandatory retirement age of seventy, but continued this work in an emeritus capacity until his death in 1977 at the age of eighty-five.

Morse was an extremely vigorous person, and at a party when he was fifty-seven, he challenged fellow mathematician Raoul Bott, more than thirty years his junior, to a race in the hundred-yard dash. Morse easily won this contest, Bott was ashamed to admit. "The overwhelming impression, which remains with me to this day, is the vast energy which Marston somehow radiated."[17]

That enthusiasm extended to the halls of academia. Bott was apprehensive before meeting with Morse during his first workday at the Institute for Advanced Study, but his uneasiness vanished when he realized that "I really did not have to say very much." Morse, Bott presumed, dominated all encounters: "I think it is a fair statement that in all conversations with Marston, one only had to do twenty percent of the talking. His energy was such that it just naturally took over."[18]

"Morse had a particular enthusiasm for his own ideas," commented Mac Lane. "There is no doubt that he liked to hear himself talk, but in my experience, Morse was a real stimulus to all who listened."[19]

One of the main jobs of those who worked for him or with him, according to Morse's former Harvard Ph.D. student Everett Pitcher, was to serve as "an audience . . . He sought collaborators and assistants, a substantial function of these individuals being to listen to his explanations of mathematical situations as he perfected his understanding of them."[20]

Maurice Heins, a Harvard mathematics graduate student in the 1930s, described collaborating with Morse as "a very intense experience. [He had] enormous drive and the physical capacity for many hours of work," said Heins. "Twenty hour days were common."[21]

Mathematics, to Morse, was a highly competitive enterprise, Pitcher said. "I . . . heard him say repeatedly that 'they' don't understand the problem. 'They' are trying to do this when it should be that. Only he understood the problem. It must be said that his position was often right."[22]

Morse was also acutely aware of priority issues in publication, Pitcher added. Once he told Morse of some research he had carried out at Morse's suggestion. Morse told him that someone else was working on the same problem, and that they ought to move quickly. "This was on a Friday," Pitcher said, "and I regarded his advice with such seriousness that I had a draft of a joint paper in his hands by Monday for him to offer to the *Proceedings of the National Academy*."[23]

For Morse, it was always a race against time—a rush to communicate as many of his ideas as he could, while he still had the chance. Stewart Cairns, another Harvard Ph.D. student of Morse's, recalled meeting with him a couple of years after his retirement in 1962. "Morse outlined to me the problems he hoped to solve, if only he could live twenty years more and keep on doing research." Morse continued working to the very end, some fifteen years into his retirement, and saw, in Cairns's words, "a substantial part of his hope fulfilled."[24]

Morse's son William also noted that up until his father's death, "he worked as fast and as hard as humanly possible to get his ideas into writing . . . In the last 10 years, it was a race against the clock and [at] about age 83 he told me he had too many ideas in his head and feared he would never get them written down, they would die with him."[25] This indefatigable drive served Morse well throughout his career. Although Morse took on a variety of topics in his career, during which he wrote some 180 papers and eight books, Morse theory was always central to his work.

Smale regards Morse's unwavering focus on Morse theory—attacking problems in the manner he was accustomed to, regardless of the techniques that others in the field were applying—as something of a mixed blessing. "What distinguished Morse was his single-minded persistence with one theme now known as Morse theory (or calculus of variations in the large)," Smale wrote in 1978. It was by virtue of that persistence, he added, "that [Morse] gave us his work of such great depth and influence." The theory he put forth constitutes an important branch in the field of global analysis, which involves the study of differential equations from a global or topological point of view. "As long as there is mathematics," Smale concluded, "Morse will be remembered with his Morse theory."[26] Or, as Cairns put it, "his fame will endure."[27]

On the other hand, one drawback of Morse's single-mindedness, according to Smale, is that it "eventually limited [his] intellectual development . . . He made no real effort to see what others were doing" and, over time, his work "became less and less relevant."[28]

Beyond just sticking to his own, tried-and-true methods, which Morse felt would allow him to make more progress in the ambitious program he had set out, he was actively resisting the infiltration of algebraic approaches into topology—or at least showing a healthy degree of skepticism. When Bott met Morse in 1949, Morse "resented the omnipresence of algebra in the topological scene at that time," Bott wrote. "The development of the algebraic tools of topology—these had little interest for him . . . For Marston always saw topology from the side of analysis, mechanics, and differential geometry," in keeping with the orientation of his mentor Birkhoff.[29]

Morse sensed that mathematics was becoming increasingly abstract, which was a shift that he embraced rather than opposed. "Mathematicians abstract in order to unify, simplify, comprehend and extend," he wrote, adding that the great complexity of modern mathematics and its vast scope make such abstractions necessary.[30] That said, Morse did not believe in abstraction for the sake of abstraction, even though he deemed it necessary in some situations. In a 1971 letter to the mathematician Arnaud Denjoy, Morse complained that "many of the young mathematicians are devising algebraic abstractions which are obvious when they are relevant and in general take more space to explain abstractly than it takes to establish the desired theorems without their use."[31]

As Morse saw it, "The battle between algebra and geometry has been waged from antiquity to the present." He took the side of geometry, resisting the influence of algebraic methods advanced by those whom he felt would like to "subordinate one [geometry] to the other [algebra]."[32]

It is fair to say that Morse was vindicated in the end through the sheer magnitude and breadth of his accomplishment. "Although Morse theory has since been recast in more modern and abstract form, its principal results have remained almost singlehandedly the work of Morse himself," Zund wrote, seventy-five years after Morse's original (1925) paper on the subject.[33]

Even so, algebraic topology proceeded on a parallel track during much of the time that Morse was developing his theory—and this emerging field produced some outstanding results, too, despite Morse's lack of enthusiasm for the subject. The basic idea of algebraic topology is to turn topological problems into algebraic ones, which can be advantageous since algebraic concepts are generally easier to work with than topological concepts. Spaces can be described through algebra and transformed through algebraic calculations, even those involving simple addition and multiplication.

Hassler Whitney

One person who contributed greatly to the success of algebraic topology was Hassler Whitney, who overlapped with Morse on the Harvard math faculty (and later on the Institute for Advanced Study faculty) and also earned his Ph.D. under Birkhoff's supervision. While both mathematicians were extraordinarily accomplished, Whitney pursued a more eclectic approach than did Morse. Garrett Birkhoff called Whitney "the most independent and self-motivated of G. D. Birkhoff's post-1930 students."[34] While that assessment is undoubtedly true, it is clearly an understatement—Whitney was one of the most independent, as well as creative, mathematicians of his generation.

Reflecting on the breadth of Whitney's work, Brown University mathematician Thomas Banchoff noted that "two things are apparent—the wide range of his interests and innovations and the solitary nature of his research. He worked almost entirely alone, although he kept up with the developments in the fields in which he initiated new ideas."[35]

Whitney's penchant for "solitude" in his mathematical research may have also guided his choice of avocation: he was a well-known mountaineer and alpinist who scaled many of Switzerland's highest peaks. In 1929, Whitney completed a first ascent on New Hampshire's Cannon Mountain on a route subsequently dubbed the Whitney-Gilman Ridge (named after him and his climbing partner, and cousin, Bradley Gilman).

The grandson of Simon Newcomb, Whitney did not show great interest in mathematics as a youth; he took few math courses in high school and none in college. But he did study physics in college, and upon reviewing his notes some time after graduating, he realized that he had forgotten most of what he had learned. "In physics it seems that you have to remember facts, so I gave it up and moved into mathematics," Whitney wrote,[36] noting that he never regretted that decision.[37]

Although his work mostly fell under the theme of algebraic topology, Whitney enjoyed moving around from subject to subject. He described his modus operandi this way: "My mathematical career consisted basically of searching out interesting problems of a simple variety; then I would plunge in... I would begin getting some results; other mathematicians would move in also, and the field would expand." Once a field started "getting too big and complex, I would move out and look for new things."[38] Or, as Whitney also put it, "when other people dove in and started building much more complex structures, I dove out."[39]

Whitney was often set off on a particular path by "little things" that might have seemed inconsequential at first. Once, for instance, when he was a National Research Council (NRC) fellow at Harvard and Princeton in the early 1930s, he ran into another Birkhoff student, Charles Morrey, who was also an NRC fellow at the time. Morrey started carrying on about "path spaces," and Whitney was having trouble following him. "I could not understand what these things were—something on a manifold, what kind of manifold I was not sure," Whitney recalled.

> I was sort of lost, and I really had things I wanted to do, so I thought, "Quick, ask him some questions." I said, "Chuck, suppose you have some kind of a curve, maybe very wiggly, in a space... How could you pick out a mid-point of that? How could you pick out a mid-point of each half? And so forth. Let's get a parameterization of it." So he started thinking about it. That enabled me to slip out. Then I started thinking about the problem myself. It tied me up for two days. I finally got an answer to it, and it turned into a nice paper of mine. [Deane] Montgomery and [Leo] Zippin quoted that paper in the book they wrote later.[40]

As Whitney described it, he sometimes latched onto significant problems only because some odd detail caught his fancy. "I tended not to be in the center of a field. When I found myself in the center of some field, I simply would not know enough of the background material." That is when he liked to move on, keepings his "eyes open for some simple elementary thing, which was elusive, to dig into."[41]

He got his Ph.D. at Harvard in 1932; his thesis was on the four-color problem, which just so happened to be one of Birkhoff's favorite topics. Whitney reformulated the problem in terms of graph theory, which led him to a noteworthy idea in linear algebra (related to the notion of "linear dependence"). That in turn motivated a great deal of research in the field of combinatorics—sometimes called the science of counting or enumeration, involving the study of possible combinations, or permutations, of the elements of sets.

While working on the four-color problem, Whitney heard from Birkhoff "that every great mathematician had studied the problem and thought at some time that he had proved the theorem. (I took it that Birkhoff included himself here.)" When Whitney was asked when the problem would be solved, he typically replied: "Not in the next half century." This prediction turned out to be pretty accurate, since the proof by Kenneth Appel and Wolfgang Haken mentioned in Chapter 3 was published in 1977, forty-five years after Whitney completed his dissertation.[42]

Although Whitney's contributions to the four-color problem and graph theory inspired important work in other areas of math, "I did not feel that that was the subject I wanted to stay in," he said, "so I was moving more into topology at that time."[43]

This turned out to be a wise move, as "Whitney was a topologist of great originality," according to the geometer Shiing-shen Chern, and "his contributions were broad."[44] By developing such concepts as sphere bundles, fiber bundles, and characteristic classes—all of which are discussed in this chapter—Whitney laid a good deal of the groundwork in algebraic topology. His timing was also propitious in that topology was still a relatively new area where American mathematicians had room to make their mark. "As a field, topology only emerged toward the end of the nineteenth century and underwent major development thereafter," writes historian Karen Hunger Parshall. "It was thus a topic that an aspiring mathematician could get in on at the ground level, and the Americans did just that."[45]

In 1935, Whitney proved that every smooth (differentiable) space or "manifold" of n dimensions can be embedded into Euclidean space of $2n + 1$ dimensions. Embedding is a mathematical way of fitting a smaller space into a larger space, although there are restrictions on exactly how this fitting is done. For example, a closed curve embedded into the larger space cannot intersect itself.

In the same year, Whitney proved a related theorem, which stated that every smooth n-dimensional space or manifold can be immersed in Euclidean space of dimension $2n$. The technique of immersion is like embedding, though somewhat less restrictive. If you take a circle and immerse it in a plane, little pieces of that circle will have no crossings. When viewed "locally," as mathematicians put it, immersion is quite similar to embedding. But if you look "globally" at the entire circle after it is immersed in the plane, there can be crossings. The curve could intersect itself once (as in a figure eight) or intersect itself two, three, four, or any number of times.

Whitney got onto this problem when Morrey had asked him how many different ways a surface bounded by a closed curve could be immersed in a familiar (two-dimensional) plane. Whitney came up with his immersion theory in the course of answering that question, and that has turned out to be a critical development in topology and geometry. Stephen Smale later generalized this problem—of immersing a manifold into Euclidean space—to higher dimensions. Mikhail Gromov, who is based at the Institut des Hautes Études Scientifiques in France and at New York University, then created a whole field called the homotopy principle, or "h principle," which he considers a powerful new approach for solving differential equations. This all grew out of a simple (though potent) idea that Whitney developed in response to a question posed by his colleague Morrey.

In 1935, Whitney also introduced the notion of "sphere spaces," which were later called sphere bundles, serving as the forerunner to the more general notion of fiber bundles, both of which are now fundamental notions in topology. The simplest example of a fiber bundle is the tangent bundle of a surface, which consists of all the tangent planes at all of the points on a smooth surface. But Whitney also did pioneering work with sphere bundles, whereby spheres (rather than planes, as in the example of a fiber bundle) are attached to every point on a surface. One strength of this idea is its generality, because the spheres that make up these bundles can be of arbitrary dimension. S^n, according to the standard nomenclature, is an n-dimensional sphere sitting in $(n + 1)$-dimensional Euclidean space. A 0-sphere consists of two points on a line, a 1-sphere is a circle in a plane, a 2-sphere is a familiar sphere sitting in three-dimensional space, and so forth.

Let us consider, for example, a circle bundle over a 2-sphere. In this case, the bundle—a circle attached to every point on an ordinary sphere—is a three-dimensional object that may, in fact, be a 3-sphere.

Simpler, though still instructive for our purposes, is a 0-sphere bundle (or two points) over a 1-sphere (or circle). The bundle over any space or manifold X assigns to every point of X something, and that something is called the fiber. We insist, moreover, that this fiber has to move continuously (without changing its topology) as we move continuously along X. The bundle thus consists of the "base" or X (a circle in this example) and the fiber (two points for each point on the circle). By sweeping out the fiber as we move along the base, we obtain the combined or "total" space, which in this case is either one circle or the union of two circles. The total space created in this way can, in general, be used to understand the topology of the original space.

Our primary interest here is not so much about 1-spheres or 2-spheres per se, but rather in the value of introducing fiber bundles in the first place, which, according to Chern, "have since become a fundamental notion of topology."[46]

One benefit is as follows: If you have a space and want to prove that it is not simply connected, you take a loop in that space and the corresponding loop in an attached bundle. If, in going around that loop once, you start at one point on the bundle and end up at another point, you will have found out that the space is not simply connected, as we did in the case of the 0-sphere bundle over the circle. The existence of this fiber bundle tells you that the circle is not simply connected. The virtue of this approach, moreover—and the reason bundles have become so important in topology—is that the same methodology works for complicated spaces that cannot be classified so readily as a 1-sphere (circle) or 2-sphere (ordinary sphere).

But Whitney carried this idea much further still, developing an essential concept linked to fiber bundles called the characteristic class. A characteristic class, in turn, is something that takes a bundle and assigns an "invariant" to it. An invariant is a topological feature or property of a space so named because it does not change even when the space itself is changed by shrinking, stretching, bending, and so forth.

The invariant in the earlier example of a 0-sphere bundle is actually a number, 0 or 1. The idea is to start with a 0-sphere bundle over a given space X and trace out a loop in X. If the loop in the bundle (which corresponds to the loop in X) takes you back to your starting point, you get a 0; if not, you get a 1. This, again, is called an invariant, because the assignment of a number (0 or 1) does not change if you transform the loop without, say, cutting it outright.

This particular invariant is called the first Stiefel-Whitney class, so named because the idea was independently, and simultaneously, developed by Whitney and the Swiss mathematician Eduard Stiefel, starting in 1935. Stiefel, however, "restricted himself to the tangent bundle," while "Whitney saw the merit of the general notion of the sphere bundle over any space," Chern reported.[47]

This is a significant distinction because there is only one tangent bundle, but there are many vector bundles. To any vector bundle you can assign Stiefel-Whitney classes. You can also assign a sphere bundle to any vector bundle. To take a simple example, a cylinder is a vector bundle of the circle, meaning that the cylinder is composed of vectors (or line segments) attached to each point of the circle. If we are talking about an open cylinder (like a can with the top and bottom removed), the sphere bundle in this case forms the boundary of the cylinder—the boundary being the two circles at the top and bottom of the can, so to speak. A sphere bundle is a much more general concept than a tangent bundle—and, as such, can assume highly varied forms—which is why it has been of fundamental importance to topology.

Whitney discovered that for each bundle one can associate invariants called characteristic classes. Simply put, Stiefel-Whitney classes assign numbers to shapes. For every non-negative integer n, there is an nth Stiefel-Whitney class; the nth Stiefel-Whitney class assigns a 0 or 1 to every n-dimensional shape within the manifold to which the bundle is attached.

The Stiefel-Whitney class is a basic tool in algebraic topology. If you have a manifold, Stiefel-Whitney classes are important invariants that can help you characterize that manifold. The first Stiefel-Whitney class has a concrete geometric meaning. It can tell you whether the space in question is "orientable" or not. A 2-sphere, such as the surface of Earth, is orientable, meaning that if two people start off at some spot on the surface, wander around, and return to that same spot, they will normally agree on which way, for instance, is clockwise and counterclockwise. A cylinder is another orientable two-dimensional manifold. A Möbius strip, on the other hand, is a nonorientable two-dimensional manifold. If people were to walk around on a Möbius strip, they could easily get turned around and have different opinions as to what is clockwise or counterclockwise.

The second Stiefel-Whitney class also has a concrete geometric meaning, which is somewhat more complicated to explain. Higher Stiefel-Whitney classes are even more abstract, though still very useful in algebraic topology.

Whitney derived a "product formula"—and proved an accompanying theorem—that shows how to combine two bundles, expressing the characteristic class of the combined bundle in terms of the original bundles. He is also credited with introducing another concept that is extremely important in topology called cohomology (although this idea has also been attributed, independently, to James Alexander and Andrey Kolmogorov at about the same time).[48] Stiefel-Whitney classes are, in fact, cohomology classes (with the first Stiefel-Whitney class being the one-dimensional cohomology class, and so forth). Indeed, the mathematician John Milnor suggested that Whitney invented the language of cohomology theory in order to establish the concept of a characteristic class.[49]

Cohomology, in essence, offered a new quantitative, as well as qualitative, way of studying topological spaces and their associated bundles. It is closely tied to the idea of homology—a topological concept dating back to Poincaré that roughly has to do with the number of "holes" that a space has. A common doughnut, or single-holed torus, has just one hole that is easy enough to see and is often remarked upon. But a topologist would say that the doughnut has two independent "cycles," or two independent loops: two independent ways of wrapping around a shape that cannot be shrunk to a point. One loop, or cycle, spans the doughnut's outer perimeter (or "equator"), whereas the second cycle winds through the hole and around the outer edge.

The so-called homology of the torus consists of those two cycles, any other elements simply being linear combinations of them. Each element of this group, in turn, forms a homology class consisting of all the cycles that are equivalent to each other. For example, if you had a rubber band going through the doughnut hole at a particular spot, shifting the rubber band sideways to a different spot would give you an equivalent cycle. Similarly, rather than having a rubber band running exactly around a doughnut's equator, you could nudge it off center or even move it to the doughnut's inner circle, and those would still be equivalent cycles.

If homology relates to the number of holes (or cycles) of different dimensions, cohomology is essentially an algebraic measure of the same thing. Cohomology encodes the same information about the space—be it a torus or some other shape—but packages it differently. Cohomology basically takes a cycle and assigns a number to it. That assignment has to satisfy some conditions, one of which is that if two cycles are equivalent and can thus be deformed into each other, then the number assigned to each cycle has to be the same. You might say that cohomology takes an n-dimensional cycle as its input and generates a number as its output.

One practical advantage of this notion, which Whitney brought into mathematics, is that cohomology lends itself to a natural form of multiplication that is more convenient than doing multiplication in homology. In fact, Whitney made use of multiplication in cohomology to express the product formula for the aforementioned Stiefel-Whitney classes. This "multiplication structure" is central to algebraic topology, and it is one reason the field has experienced such great advances—many of which owe to innovations made by Whitney.

Whitney presented what he called "a fairly full account" of the research on sphere bundles in a 1941 paper. He planned to write a more comprehensive treatment of the subject in his own book, but he first wrote another book that was published in 1957, *Geometric Integration Theory*, which he thought would lay the foundations for the sphere bundle book.[50] By this time, he was well established at the Institute for Advanced Study, having moved from Harvard to Princeton in 1952. A mathematics colleague, Warren Ambrose, suggested to Whitney around this time "that the publication date of your book on sphere bundles is receding at the rate of two years per year." Whitney agreed that the process was not converging and, consequently, never got around to writing that book. But he continued to write additional papers, about which he was "amply satisfied," sometimes venturing outside his normal areas of work.[51]

He shifted gears again in the late 1960s, moving away from research and focusing his efforts instead on elementary school mathematics education. When Whitney died in 1989, his ashes were spread at the summit of Dent Blanche, a Swiss mountain he had climbed with his cousin seventy years earlier.

In a survey of his field of mathematics, which he completed shortly before his death, Whitney discussed how algebraic topology had come of age: "I have seen general topological and algebraic methods flourishing all over the world increasingly as the 'center of mass' of such studies moves still nearer to the U.S.A. shores."[52]

By the late 1930s, wrote the mathematician Alex Heller, "algebraic topology had amassed a stock of problems, which its then available tools were unable to attack. A small group of mathematicians . . . dedicated themselves to building a more adequate armamentarium."[53] Whitney was prominent among this group, but so was Saunders Mac Lane, a Harvard colleague who helped lead the charge, building directly upon Whitney's work on cohomology and characteristic class, while forging some new mathematical directions as well.

Saunders Mac Lane

Although he eventually assumed a highly visible leadership position in his field, Mac Lane took a rather indirect route in getting there. In his early years as a Yale undergraduate, Mac Lane recalled, "I was not yet aware that there was such a thing as a career in mathematics." Nor was he "aware of [mathematical] research that led to new results." While he found calculus exciting, he said, "it seemed as though it had long since been entirely worked out."[54]

Out of deference to a prosperous uncle, who paid for his college education and hoped his nephew might go into business or law, Mac Lane tried an accounting course but found it excruciatingly dull. He also tried a chemistry course, a field in which he assumed he could find a job with sufficient compensation, but that struck him as dull, too. It was not until he met a young assistant professor, Øystein Ore, who had recently come to Yale from Norway, that Mac Lane realized "there were brand new ideas to be found in mathematics. With this indication, my focus shifted from the accumulation of knowledge to the hope of discovering new knowledge . . . I could see that there were many new things to be done."[55]

After graduating from Yale in 1930, Mac Lane went to the University of Chicago for graduate school, mainly because he was offered a scholarship. But he left after a year, disappointed because he did not see any possibility of pursuing his interest in mathematical logic. He went instead to Göttingen, which was strong in that area and still widely regarded as the world's leading mathematics department at the time. Arriving in Göttingen in 1931, Mac Lane was exposed to—as well as inspired by—modern algebra as practiced by Emmy Noether, whom he considered "probably the most important woman mathematician of all time."[56]

The Nazis came to power in 1933, and most faculty members with Jewish ties—at Göttingen and at other universities—were summarily dismissed. Paul Bernays, Mac Lane's thesis supervisor, was forced to leave the school as a result, along with Noether and many others. Mac Lane arranged for Hermann Weyl to oversee his Ph.D. work. Weyl had once said, in a quote reminiscent of Marston Morse: "In these days the angel of topology and the devil of abstract algebra fight for the soul of every individual discipline of mathematics."[57] Weyl may not have realized that his new advisee might soon bring that angel and devil together into one neat package.

With the situation rapidly deteriorating at Göttingen and throughout Germany, Mac Lane resolved to finish up his dissertation (on the subject of mathematical logic) as soon as possible. He defended his thesis in July 1933, with Weyl serving as examiner, and promptly returned to the United States. Reflecting on his turbulent years in Germany, Mac Lane often thought about the time he went to the opera house and found himself standing next to Hitler and Goebbels during the intermission. Afterward, he wondered how history might have gone differently if he had had a gun with him at that time and used it. "It thus later seemed to me to be the one occasion where (had I carried a weapon) I might have personally changed history."[58]

After returning to America in 1933, Mac Lane attended an American Mathematical Society meeting in Cambridge, Massachusetts, where he spoke with George D. Birkhoff about a job. He also interviewed at Exeter, a prep school in New Hampshire, regarding an opening for a mathematics "master." In view of the dire economic climate, with the Great Depression still in full swing, Mac Lane had to give serious consideration to a high school teaching position—albeit at one of the nation's most prestigious high schools. "It is interesting to contemplate what course his career might have taken if this leading American mathematician had gone to Exeter instead of Harvard," wrote the mathematician Ivan M. Niven.[59]

But Mac Lane was spared that difficult choice when, within a couple of weeks, the Harvard mathematics chair, William Caspar Graustein, offered him a two-year appointment as a Benjamin Peirce Instructor, asking him to teach an advanced course. "It's clear that I profited from Birkhoff's policy" of taking care of young American mathematicians rather than giving priority to established scholars emigrating from Europe, Mac Lane wrote. Needless to say, he accepted Graustein's offer. "It meant an adequate salary, and a chance to give advanced courses and to talk to George Birkhoff and Marston Morse. There were some awfully good people at Harvard at that time."[60]

At Harvard, Mac Lane offered to teach a course on mathematical logic but was encouraged to teach algebra instead, since at that time there was no algebraist on the Harvard faculty. He taught the modern, abstract version of algebra championed by Noether and imported directly from Germany. After two years at Harvard, he taught for a year at Cornell and a year at Chicago, before returning to Harvard in the fall of 1938 as an assistant professor. Mac Lane stayed at Harvard until 1947, rising to the rank of full professor. It was, he said, "a fine time to be in a first-class department at a leading university."[61]

Mac Lane shared algebra instruction chores with Garrett Birkhoff, who also became an assistant professor at Harvard in 1938. Birkhoff had already spent a number of years at the university, getting his undergraduate degree there, serving as an instructor, and becoming a junior fellow from 1933 to 1936 in Harvard's newly formed Society of Fellows, which Mac Lane described as "a place for advanced students who couldn't be bothered to get a Ph.D."[62] The society was modeled after a similar program at Trinity College in Cambridge, England, that the mathematician and philosopher Alfred North Whitehead had participated in. After arriving at Harvard in 1924, Whitehead suggested that Harvard start a similar program—an idea that appealed to the school's president at the time, Abbott Lawrence Lowell.[63] (Since its founding in 1933, the society has been home to many mathematicians in addition to Garrett Birkhoff, including Stanislaw Ulam, Andrew Gleason, Lynn Loomis, David Mumford, Barry Mazur, Robin Hartshorne, Clifford Taubes, Noam Elkies, and Dennis Gaitsgory.)

Birkhoff was a vital contributor in the 1930s to the field of "universal algebra" (as was Øystein Ore, Mac Lane's former teacher). Universal algebra is an abstract approach to algebra that involves, for instance, the study of the theory of groups in general rather than the study of individual groups and their specific properties. A group is an almost ubiquitous structure in mathematics—a set of elements (which could be numbers, other objects, or even other sets) that satisfy specific rules: Each group has an identity element (such as the number 1) and an inverse element (such as $1/x$ for every x). A group is "closed," meaning that when two elements are combined through an allowed operation (such as addition or multiplication), the result will always be a member of the group. Furthermore, these operations must always obey the associative law: $a \times (b \times c) = (a \times b) \times c$. This is an admittedly brief introduction to group theory, though it is worth pointing out that the definitions of a group may be recast a bit for the purposes of universal algebra.

Birkhoff proved a theorem, which is now named after him, "that there could be a 'real' universal algebra," Mac Lane said, which concerns the properties common to all algebras and all algebraic structures (such as groups, fields, and so forth). "Birkhoff's theorem . . . became the starting point for the subsequent active development of universal algebra."[64]

Birkhoff and Mac Lane taught undergraduate algebra in alternating years and combined their presentations to produce a joint book, *Survey of Modern Algebra,* which was the first undergraduate algebra textbook in the country that introduced the abstract notions of Emmy Noether. Birkhoff and Mac Lane were both promoted from assistant to associate professor in 1941—the same year that their book was published. "It is unfortunate that in today's climate young mathematicians might not consider writing such texts out of concern that the time taken from their research would negatively affect their chances for promotion," Mac Lane wrote, although this book, evidently, did not hurt their chances.[65]

It did, however, help the chances of many mathematicians, young and old, seeking to grasp the latest thinking in algebra. Gerald L. Alexanderson of Santa Clara University calls the book "one of the most famous mathematical texts in English, one from which most mathematicians of a certain generation learned their abstract algebra."[66] *Modern Algebra* was "incredibly influential," adds Harvard's Barry Mazur. "When I was a kid, that was *the* book. There was no competition."[67]

For Mac Lane, 1941 was momentous for another reason: in that year he began an extremely fruitful collaboration with Samuel Eilenberg, a topologist who had left Poland two years earlier with the threat of war imminent. Thanks to the help of Veblen and Solomon Lefschetz at Princeton, who tried to find places for mathematical refugees from Europe, Eilenberg (also known as S^2P^2, "Smart Sammy the Polish Prodigy")[68] had landed a position at the University of Michigan, which was then strong in topology. Mac Lane had been invited to give six lectures at Ann Arbor about "group extensions," which is a way of combining groups and building bigger groups out of smaller ones. Eilenberg attended the first five lectures but had to miss the sixth, so he asked Mac Lane if he would discuss the talk with him afterward. Mac Lane was happy to comply.

While conferring later that night, Eilenberg noticed that Mac Lane's group calculation, which was rooted in pure algebra, was strikingly similar if not identical to a calculation in topology that was seemingly unrelated. The topological calculation concerned the homology (or "cycles"), as well as the cohomology, of a strange, infinitely twisted shape called the p-adic solenoid (where p can be any prime number and "p-adic" refers to a novel kind of numbering system, introduced more than a century ago, with important applications in number theory). The solenoid can be made, according to Mac Lane, "by taking a torus [or doughnut], wrapping a second solid torus p times around inside the first torus, wrapping a third torus p times around inside the second one, and so on," up to infinity.[69] Why on earth, they wondered, did two calculations with such different motivations and origins look virtually indistinguishable? "The coincidence was highly mysterious," Mac Lane said. They stayed up all night trying to get to the bottom of it, and by morning they had some idea of what was going on.[70] But it would take years, for them and others, to sort through the implications of "this unexpected connection between algebra and topology," as Mac Lane put it.[71]

The collaboration became easier in 1943, when Mac Lane took a leave from Harvard and went to Columbia (where Eilenberg taught) to direct the Applied Mathematics Group during World War II. The group was charged with helping the U.S. Air Force solve technical problems such as figuring out how Allied bombers should target German fighter planes that were gunning for them. Morse and Whitney worked with Mac Lane on these problems, as did Irving Kaplansky (Mac Lane's first Ph.D. student at Harvard), the topologist Paul Smith, George Mackey (a Harvard math Ph.D. who would soon join Harvard's permanent faculty), and Eilenberg. Mac Lane worked for the Air Force during the day in what was, for him, a rare foray into applied mathematics. At night, he and Eilenberg immersed themselves in pure mathematics, leaving little time for anything else.

As a team, Mac Lane and Eilenberg complemented each other well. "It so happened that this was a time when more sophisticated algebraic techniques were coming into algebraic topology," Mac Lane noted. "Sammy knew much more than I did about the topological background, but I knew about the algebraic techniques and had practice in elaborate algebraic calculations. So our talents fitted together."[72]

What they came up with in the end went far beyond understanding just one particular coincidence. Eilenberg and Mac Lane showed how the algebraic problem involving group extensions and the topological problem involving the p-adic solenoid were one and the same by translating each problem into a common language that employed new, abstract terms like categories, functors, and natural transformations. When the two problems were expressed using the same terminology, it was hard to miss the fact that they were identical. In the process of investigating this unexpected connection, and in explaining how it had come about, they ended up inventing category theory, which provided a uniform framework showing how different branches of mathematics and different mathematical formulations relate to one another.

Eilenberg and Mac Lane could have explained the link between group extensions and the solenoid without inventing category theory, says Michael Barr of McGill University. "But they wanted to explain where this particular commonality came from. They were trying to find a more principled explanation, and that more principled explanation led to category theory."[73]

The authors consistently underestimated the importance of the theory they jokingly referred to as "general abstract nonsense." Of course, Mac Lane noted, "we didn't really mean the nonsense part and we were proud of its generality."[74] Eilenberg, for example, believed their first comprehensive paper on category theory, which was published in 1945, would be the only paper ever needed on the subject. Even getting that one paper published was a bit of a challenge (though made easier by the fact that both Mac Lane and Eilenberg were by then well-known mathematicians) because the paper was so abstract that some of their peers considered it wholly lacking in content.

But in a way, that was the point. The mathematician Norman Steenrod, who was then based at the University of Michigan (before moving to Princeton University) said that their 1945 paper "had a more significant impact on him than any other research paper; other papers contributed results while this paper changed his way of thinking."[75] The paper, which came out in the *Transactions of the American Mathematical Society,* was somewhat of an inside job, in that the *Transactions* editor was Paul Smith, who was then in the Applied Mathematics Group, and the paper was refereed by George Mackey, who was also a member of the group. Not that Eilenberg and Mac Lane needed to cut corners—their paper turned out to be one of the most influential of its time.

But Eilenberg's prediction of one paper being sufficient to cover category theory turned out to be way off the mark, as category theory has proved to be far richer than that. Mac Lane estimated that in the mid-1960s, nearly twenty years later, more than sixty mathematicians started work in this area, based on his unofficial tally of the first research papers published on the subject.[76]

Category theory, according to Mazur, "set the stage for all modern unifications of mathematics. So many theories were begging for it, and it is now the lingua franca."[77] But the theory does more than just show how different mathematical subdisciplines relate to one another. "The language they [Eilenberg and Mac Lane] introduced there transformed modern mathematics," said Peter May of the University of Chicago. "In fact, a very great deal of mathematics since then would quite literally have been unthinkable without that language."[78] Steve Awodey of Carnegie Mellon University agrees, saying of category theory that "today wide swaths of mathematics cannot even be formulated without it."[79]

Category theory has also contributed to the proofs of key theorems, even though the theory's founders did not originally envision such an outcome. Alexander Grothendieck, for example, used tools from category theory to prove conjectures posed by André Weil, which were statements of great consequence in algebraic geometry. Without the advent of category theory, and the new avenues it provided, it is not clear how or whether Weil's conjectures would have been proved.

Whereas William Lawvere, a former graduate student of Eilenberg's, saw category theory as the foundation of all mathematics, Eilenberg and Mac Lane were much more conservative in their assessment. But over time, Mac Lane came around to Lawvere's position on many points. At his eighty-fifth birthday party, Mac Lane asked a colleague: "Do you think I've done everything I can to promote Lawvere's ideas?"[80]

Eilenberg and Mac Lane wrote fifteen joint papers in the course of about as many years—work of greater moment than anything either of them had done alone. In addition to inventing category theory, which vindicated Mac Lane's earlier investment in logic, they are also famous for the topological spaces that bear their names.

Eilenberg-Mac Lane spaces are another profound implication of the strange link between algebra and topology that Eilenberg first noticed in 1941. The problem that Mac Lane was working on at that time involved the algebraic definition of the cohomology of groups—a calculation that did not involve topology at all. But through the connection observed by Eilenberg, they were able to see that the group cohomology worked out by Mac Lane is exactly equal to the cohomology of the topological space that can be constructed for that group. Such a space is a special case of what are now called Eilenberg-Mac Lane spaces.

Eilenberg-Mac Lane spaces have a special property that relates to the concept of homotopy. Two topological spaces are considered to be "homotopy equivalent" if one can be continuously deformed to the other. A cylinder and a circle are examples of homotopy equivalent spaces, because you can get a circle by squashing a cylinder—a so-called many-to-one mapping in which many points on the cylinder are carried to a single point on the circle. This stands in contrast to the more restrictive concept of a homeomorphism, which involves a continuous, one-to-one mapping between two topologically equivalent spaces.

With that as a background, we will now try to explain what makes Eilenberg-Mac Lane spaces so special. One of their defining features is that if you map from a sphere onto such a space, all the maps will be "trivial," in the sense that they will all take you to the same point on the Eilenberg-Mac Lane space—unless you pick a sphere of the right dimension n, where n is a positive integer. Only one choice of n will lead to an interesting (nontrivial) map; all other possible choices of n will lead to a boring map that takes everything—every point—to the same place.

Another way of putting it is that an Eilenberg-Mac Lane space has nonzero homotopy in only a single dimension. That is a desirable feature, explains Colin McLarty of Case Western Reserve University, because a complicated space that has nonzero homotopy (or, more technically, a nonzero homotopy group) in finitely many dimensions can be built up from simpler spaces that have nonzero homotopy in just a single dimension.[81]

And that, in essence, is why Eilenberg-Mac Lane spaces have proven to be so useful in algebraic topology. They are building blocks out of which you can make arbitrary spaces. For many problems, if you understand the building blocks—the Eilenberg-Mac Lane spaces—then you can understand the more general case, too. This approach has indeed been widely embraced, notes Lawvere: "Much of algebraic topology in the 1950s and 60s and beyond is built upon the idea of Eilenberg-Mac Lane spaces."[82]

Mac Lane continued to pursue these ideas and to explore the spaces named after Eilenberg and him, but by this time he had moved from Harvard to the University of Chicago. Marshall Stone, who was charged with rebuilding Chicago's math department, persuaded Mac Lane to accept a position there in 1947. Mac Lane remained at Chicago for the rest of his career, which lasted more than another half century. One reason Mac Lane offered to explain the move from Harvard was that his wife, Dorothy, "never fit in well with the traditions or autocracy in New England."[83] Another reason was that in Stone's effort to bolster the university's math department—drawing on talented refugee mathematicians to do so—Stone was trying to advance the nation's mathematics as a whole. Mac Lane decided to join him in this endeavor, as he, too, was a tireless champion of American math.

In the latter half of his long career, Mac Lane became highly visible on the national mathematics scene. When Stone stepped down as the department head in 1952, Mac Lane became the chair for six years. In time, he took on more administrative duties, becoming president of the American Mathematical Society and the Mathematical Association of America, vice president of both the National Academy of Sciences and the American Philosophical Society, a member of the National Science Board, and an adviser to various government committees. When he assumed the presidency of the American Mathematical Society, states Della Fenster of the University of Richmond, Mac Lane helped turn what had been largely a figurehead position into more of an advocate for mathematics.[84]

Mac Lane was devoted not only to mathematical research and to promoting his field but also to mathematical instruction. He supervised forty Ph.D. students overall, including John Thompson, an eventual Fields Medal winner; Kaplansky, who became Chicago mathematics department chair and later became director of the Mathematical Sciences Research Institute in Berkeley; and David Eisenbud, who headed that institute as well.

Through these activities and through his published work, Mac Lane certainly earned his nickname of "Mr. Mathematics."[85] He was considered an "outspoken mathematician" of "uncompromising principles," who never refrained from expressing his opinion or from exerting his will. Walter Tholen, a colleague of Mac Lane's who is now based at York University, is not likely to forget an incident that took place in the early 1970s at the Mathematical Research Institute in Oberwolfach, Germany, where a group of mathematicians were taking an afternoon hike. "John Gray [a category theorist based at the University of Illinois] was heading in one direction, and Saunders was pointing his cane in the other, while vehemently disputing John's claim about the right direction," Tholen recalled. "Of course, neither of the two men retreated, leaving everybody else with the difficult decision of whether to follow what would most likely be the better direction, or to simply follow the boss. With most people choosing the latter option, only a few arrived back in time for dinner at the Institute."[86] While Mac Lane might have led his peers onto a divergent path in this one instance, during the bulk of his career he helped forge a great unification in mathematics that has yet to run its course.

5

ANALYSIS MOST COMPLEX: LARS AHLFORS GIVES FUNCTION THEORY A GEOMETRIC SPIN

Born in 1907 in Helsinki, Finland, Lars Ahlfors was the first European mathematician to hold a permanent position on the Harvard mathematics faculty. In deviating from its traditional policy of recruiting "homegrown" American mathematicians, the department was hardly taking a chance in 1946 when it hired Ahlfors (following a three-year visiting appointment from 1935 to 1938), since by that time he had already established himself as an international star of the first rank. In 1936, Ahlfors was the winner (along with Jesse Douglas of MIT) of the first Fields Medal ever offered—a prize that has since come to be regarded as mathematics' highest honor. Years later, he was granted other prestigious awards, including the Wolf Prize in Mathematics and the Steele Prize.

Throughout his career, Ahlfors made deep, seminal contributions in many distinct though related parts of mathematics. Most of this work fell under the heading of complex analysis—also known as complex function theory—the study of functions whose variables consist of complex numbers. The brilliant nineteenth-century mathematician Bernhard Riemann originated many of the techniques for using geometry to study complex analysis and, conversely, using complex analysis to study geometry. Ahlfors took up this approach in the twentieth century, making one key finding after another—a number of which are discussed in this chapter. As a leader in this area, Ahlfors often set the agenda for the entire field.

"Ahlfors was *the* great pioneer in our lifetime of complex analysis, especially from a geometric perspective," said University of Minnesota mathematician Albert Marden, a former Ph.D. student of Ahlfors's at Harvard. "One could argue that Ahlfors defined the field [of complex analysis] by the scope of his work."[1] He focused, in particular, on the theory of Riemann surfaces—one-dimensional complex spaces (or "manifolds") or two-dimensional real surfaces—which provided a geometric way of looking at functions of a complex variable.

On numerous occasions, advances made by Ahlfors opened up fertile areas of mathematical research. A Finnish colleague, Olli Lehto, described Ahlfors as "a King Midas: Whatever he touched always turned to gold."[2]

None of this, of course, could necessarily have been foretold in Ahlfors's youth, during which, by his own admission, he exhibited no signs of being a child prodigy. He was "fascinated by mathematics without understanding what it was about," Ahlfors acknowledges, having had no access to any mathematical literature until he reached the highest grades.[3] His mother died during his birth, and he was raised by his father, a professor of mechanical engineering, who was "very stern," according to Ahlfors.[4] "Having seen many prodigies spoiled by ambitious parents, I can only be thankful to my father for his restraint. The high school curriculum did not include any calculus, but I finally managed to learn some on my own, thanks to clandestine visits to my father's engineering library" at their home.[5]

As a child Ahlfors was a bookworm who shunned many of the activities that boys his age typically engaged in. He disliked sports and hiking in the mountains and "hated vacations and Sundays, for I had nothing to do on those days," he said. He also had no passion for history. "Why memorize the years associated with various events and people? One might just as well memorize telephone numbers. It didn't make any sense to me, and it seemed rather silly. My history teacher was not very fond of me."[6]

But Ahlfors never tired of mathematics and enjoyed working on problems in his spare time. He was absolutely delighted, his friend Troels Jorgensen recalls, "when, as a young man, he found out that one could become a mathematician."[7] His father supported that proposition, especially after realizing that his son was not cut out for engineering. "He had wanted me to become an engineer, like himself, but had soon found out that I could not do anything mechanical, not even put in a screw," Ahlfors said. "So he decided that I should go into mathematics."[8]

Ahlfors entered the University of Helsinki in 1924, where he studied under Ernst Lindelöf and Rolf Nevanlinna. Lindelöf, the reigning patriarch of Finnish mathematics, had been the only full professor of mathematics at the university until Nevanlinna, a leading light in the mathematical world, was hired.

As a freshman, Ahlfors brashly forced his way into Nevanlinna's advanced calculus course. Saying that he lacked the prerequisites would have been an understatement. In fact, he had not taken a single calculus course before, other than what he had picked up on his own. "I persuaded [Nevanlinna] to let me stay," Ahlfors wrote. "Later I found out that Lindelöf disapproved of this rash decision of his younger colleague. I was too young, and I am afraid I was a pest." Nevanlinna, however, was a "superb teacher," in Ahlfors's estimation. "It was, of course, preordained that I should specialize in [complex] function theory" or complex analysis, as it is also called.[9]

Lindelöf had, in a sense, set the agenda for much of mathematics throughout the country. "In the 1920s, all Finnish mathematicians were his [Lindelöf's] student," Ahlfors said. "Because of Lindelöf, complex analysis had . . . become a special subject for Finnish mathematicians." Ahlfors took up the subject himself during his second year of studies.

In 1928, the year after he graduated, Ahlfors accompanied Nevanlinna to the Federal Polytechnic Institute in Zurich. Nevanlinna was replacing Hermann Weyl, who was on leave during the 1928–29 academic year. Lindelöf took Ahlfors aside and told him that he must somehow find a way to get himself to Zurich with Nevanlinna, no matter what, "even if you have to go dog-class."[10]

The year spent in Zurich was a revelation for Ahlfors. "This was my first exposure to live mathematics, and it opened my eyes to serious thinking as opposed to passive reading," Ahlfors said.[11] It was during this year that he came to understand what mathematics was about—"that I was supposed to do mathematics, not just learn it. This had not been clear to me before." For the first time, he was exposed to unsolved problems in mathematics, which got him excited, motivating him to go further and, he hoped, come up with something new—something no one else had done before.[12]

In Nevanlinna's course on contemporary function theory, which Ahlfors attended, Nevanlinna introduced his class to Denjoy's conjecture, which concerned the number of asymptotic values of an entire function. The conjecture held that an entire function of order k can have at most $2k$ finite asymptotic values. Since that terminology is a bit imposing for the nonspecialist, let us take a moment to explain what this conjecture is about. An entire function, such as $f(z)$, where $z = x + iy$, is a function that is differentiable over the whole complex plane, which is a plane represented by a real (x) axis and an orthogonal imaginary (iy) axis. Examples of entire functions include polynomial functions (like $ax^2 - bx + c$), which consist of variables (such as x and y) and constants (such as a, b, and c) that can be added, subtracted, and multiplied together, using only positive integer exponents; exponential functions (of the form e^z); and trigonometric functions (such as the sine and cosine); as well as sums and products of these functions.

Roughly speaking, the order of an entire function—similar to the degree of a polynomial—tells you how fast the function grows, and the order does not have to be finite. If the entire function is a polynomial, its order, by definition, is zero. An exponential function of the form e^z is said to be of order one. A function of the form e^{e^z} is of infinite order. While these values—order zero, one, or infinity—form a basis for comparison, the order of entire functions, in general, can be derived from a somewhat complicated expression, which we do not go into here, that involves taking the limit of the log of the function inside a disk on the plane as the radius of that disk goes to infinity.

Denjoy's conjecture, in some sense, tries to "get a handle" on a function that traces out a path in the plane, placing some constraints on that function when possible. In this case, the path that is traced out need not be a straight line. It could wiggle around, heading off to infinity in some directions, and it could be bounded in other directions, where it approaches some definite number.

By starting at the origin and drawing rays that divide the plane into sectors, one may find that the function is bounded in some sectors but not in others. In sectors where the function is bounded, it may oscillate around, never settling down to a particular value. Or it may converge toward some finite, "asymptotic" value. (For example, the function e^x goes to zero when x approaches negative infinity.) According to Denjoy's conjecture, the number of distinct asymptotic values for this function is at most $2k$, though it could be less.

Upon being introduced to this problem by Nevanlinna, Ahlfors was immediately transfixed. "From that point on, I knew I was a mathematician," Ahlfors said. "What could have been more tantalizing than to hear of a problem that seemed approachable and had been open for twenty-one years, precisely my age?"[13]

Ahlfors came up with a new way of attacking the problem. His approach was based on "conformal mapping," which involves a function that preserves angles. He also relied on the so-called area-length method, which has to do—as the name implies—with the relation between length and area. The disk offers perhaps the simplest example. For a very small disk, and hence a very small radius (r much less than one), the length of the bounding circle ($2\pi r$) is large compared with the area of the circle (πr^2). Conversely, for a very large disk and a large r (r much greater than one), the area is large compared with the length. Ahlfors exploited this same general relation between length and area in the more involved setting of Denjoy's conjecture. "If you consider this problem of asymptotic values, there is absolutely no mention of length and area, and it's not obvious that you should do anything geometric at all to solve it," says University of Illinois mathematician Aimo Hinkkanen. "But if you take a geometric approach, you just might be able to solve the problem, and that's what Ahlfors did."[14]

"I had the incredible luck of hitting upon a new approach, based on conformal mapping, which with very considerable help from Nevanlinna and [George] Pólya led to a proof of the full conjecture," Ahlfors wrote. "With unparalleled generosity, they forbade me to mention the part they had played... For my part, I have tried to repay my debt by never accepting to appear as coauthor with a student."[15]

Ahlfors's proof was published in 1930 as part of his doctoral thesis, which expanded on this result. The proof, which had eluded many of the leading mathematicians of the day, secured his fame at an early age. Although Ahlfors did not consider the proof so important on its own, he admitted that "it was new and it showed the way to other things."[16] It also transformed a hitherto unknown twenty-one-year-old into something of a mathematical celebrity.

Ahlfors then followed Nevanlinna to Paris for three months. There he met Arnaud Denjoy, who told him that "21 is the most beautiful number, because 21 years ago he had made this conjecture, and when I was 21 I had solved the problem."[17]

Soon afterward, a young Swedish mathematician named Arne Beurling, whom Ahlfors had not yet met, came up with his own proof of the conjecture. "It is not unusual that the same mathematical idea will surface, independently, in several places, when the time is ripe," Ahlfors later wrote. "My habits at the time did not include regular checking of the periodicals, and I was not aware that [the German mathematician Herbert] Grötzsch had published papers based on ideas similar to mine, which he too could have used to prove the Denjoy conjecture. Neither could I have known that Arne Beurling had found a different proof in 1929 while hunting alligators in Panama . . . It is interesting that we all used essentially the same distortion theorem for conformal mapping."[18]

Upon returning to Finland, Ahlfors took his first teaching position as a lecturer at the Swedish-language university Åbo Akademi in Turku, Finland. (Although he was born in Finland, Ahlfors's native language is Swedish, as he grew up among the minority of Swedish-speaking Finns.) He became an associate professor at the University of Helsinki in 1933. He got married that same year to Erna Lehnert, describing his wedding as "the happiest and most important event in my life."[19] By this time, Ahlfors's research career was in full swing.

"The solution of Denjoy's conjecture was the prelude to Ahlfors's exceptionally fruitful mathematical production," Lehto wrote. "In the 1930s, Ahlfors achieved many significant results in complex analysis by combining standard methods with new ideas from differential geometry and geometrical topology."[20]

Ahlfors's work was "infused with a deep geometric sense," added Stanford mathematician Robert Osserman, who was Ahlfors's graduate student at Harvard in the 1950s. Rather than directly contributing to the field of geometry itself, Osserman said, Ahlfors instead made "brilliant use of differential geometry in various parts of function theory."[21]

In 1935, Ahlfors came to Harvard on a three-year appointment as a visiting lecturer, based on the recommendation of the influential German mathematician Constantin Carathéodory, who told the department chair William Caspar Graustein that "there is a young Finn . . . , and you should try to get him."[22] Ahlfors, who was already conversant in Swedish, Finnish, German, and French, had somehow managed to pick up some English prior to his arrival in Cambridge. When asked when and how he had learned the language, Ahlfors replied: "I came by boat."[23]

Ahlfors published two influential papers in 1935, each of which provided his own geometric version of an aspect of "Nevanlinna theory" (also called value distribution theory). This theory was regarded as one of the key areas of mathematical research in the 1930s, so it made sense for Ahlfors to begin his work in this subject, especially given that Nevanlinna was his former professor and mentor, as well as a friend.

The starting point for much of Nevanlinna theory was Picard's theorem, which was proved in 1879 by the French mathematician Charles Émile Picard. The first part of the theorem, sometimes called "Little Picard," states that if you have an entire (as in differentiable) function in the complex plane that does not take on (or "omits") two or more finite values, then that function must be constant, meaning that it assumes the same value everywhere on the plane. Putting this another way, if a complex "analytic" (or infinitely differentiable) function, $f(z)$, is entire and nonconstant, it can assume any value on the finite plane except one. (For example, the exponential function $f(z) = e^z$ can assume any value except zero.) "Picard's proof was sort of a miracle, like an inspiration from God, which worked without really explaining anything," notes Purdue mathematician David Drasin.[24] "The proof, while extremely short, gives no insight on why it is true and also yields no additional information."[25]

By drawing on basic principles, Nevanlinna, on the other hand, found a way to recast Picard's theorem, reformulate it, and prove it in a completely different manner. In papers he wrote in the mid-1920s, Nevanlinna developed a much more general method for relating the number of solutions to an equation, such as $f(z) = t$, to the properties of the function itself. He framed the problem by drawing a circle of radius r on the plane, where the absolute value of z is less than r, and figuring out how many solutions lie within that circle. (This approach, explains Drasin, is in the spirit of the argument that if f is a polynomial of degree n in the complex plane, then the equation $f(x) = a$ is normally expected to have n solutions or "roots.")[26]

Ahlfors continued the work of his mentor, but while Nevanlinna theory was firmly rooted in the techniques of analysis or calculus, Ahlfors obtained a geometric understanding of Picard's theorem and of Nevanlinna's main results without relying at all on analytic function theory. Rather than focusing on the number of solutions lying at a given point—or "covering" a given point—as Nevanlinna did, Ahlfors wanted to see how many times you cover a given *region*. His work constituted a geometric generalization of Nevanlinna theory, which was itself a generalization of Picard's theorem. "Ahlfors's theory can do things that Nevanlinna's theory cannot do and vice versa," says Hinkkanen, adding that "Ahlfors's success with the Denjoy conjecture may have given him the motivation and confidence to follow his geometric insights."[27]

Ahlfors's great innovation, alluded to above, involved covering surfaces. In particular, he considered functions that map the complex plane to a sphere. The approach could be used to map the plane to other Riemann surfaces, but in introducing this complicated technique, it was reasonable for him to start with one of the simplest cases. Nevanlinna theory, which Ahlfors was building upon, was also based on mapping from the complex plane to the sphere. To get a simplified picture of what Ahlfors was up to, let us take an ordinary plane in three-dimensional space and place a sphere on top of that plane, centered at the origin. The sphere's south pole intersects the plane at the origin; the north pole lies exactly on top of the sphere. If you draw a straight line from any point on the plane to the north pole, it will intersect a point on the sphere. In fact, by repeating this process from every point on the plane, you can map the plane to the sphere, getting the whole sphere except for one point, the north pole, which to a mathematician represents infinity. In this case, the mapping or function misses just a single point on the sphere, which is exactly the kind of problem that Picard's theorem addresses.

Ahlfors and Nevanlinna were interested in a special class of functions, called meromorphic, which consist of the quotient of two holomorphic (or "analytic") functions—the latter being "well behaved," as in differentiable, functions of a complex variable. A meromorphic function is itself well behaved, except at a set of isolated points where the denominator of the fraction is zero. Ahlfors showed that a meromorphic function that mapped a complex plane into a sphere would miss, at most, two points. He showed, in other words, that the maximum number of values omitted—two in this case—is the same as the Euler characteristic of a sphere. This is not coincidental, because the Euler characteristic, X, of a surface relates to the genus of a surface (g), according to the formula $X = 2 - 2g$. An ordinary sphere, having no holes, has a genus of 0, and hence its Euler characteristic is 2. The Euler characteristic comes into Ahlfors's analysis because a sphere that is missing some points—and is no longer smooth in the vicinity of those points—is no longer a perfect sphere. It is a modified sphere, and its Euler characteristic may be modified as well.

A rational function (the quotient of two polynomials) of degree d will cover the sphere d times. An arbitrary meromorphic function, which is not rational, might cover the sphere an infinite number of times. Since that is not ideal for counting, for practical reasons one might instead consider that the domain or input for the function be restricted to just a portion of the complex plane—say, a disk of radius r. The output of that function, in turn, will cover a portion of the sphere a finite number of times. As the radius gets larger and larger, of course, the sphere gets covered more times.

While the function will not necessarily cover every individual point on the sphere the same number of times, Ahlfors showed that a region or domain on the sphere will essentially be covered the same number of times, and this holds for every part of the sphere. He found, moreover, that for rational functions each section of the sphere is covered exactly the same number of times, whereas for meromorphic functions this is more or less true, albeit with some minor variation.

Many of Ahlfors's findings were summed up in a complicated inequality—sometimes called his "main theorem of covering surfaces"—that involves the Euler characteristic of the modified sphere (which may be missing some points) and the length and area of the circle bounding the disk on the plane. As with his proof of Denjoy's conjecture, the length-area relation is again critical in his proof of covering surfaces.

Ahlfors's work on covering surfaces, encapsulated in a 1935 paper published in *Acta Mathematica*,[28] was "deep and fundamental," according to Hinkkanen, "and even today it stands as one of the deepest results in mathematics. So far, no simplification has been found. If you want to construct an argument like this, you have to go through all the steps that Ahlfors went through."[29]

Not many of Ahlfors's contemporaries did that, perhaps because the protocol for covering surfaces is rather difficult to do even though the idea, on a conceptual level, may seem rather simple. Initially, Ahlfors did not have any of his students pursue the subject further because he was just a temporary professor at Harvard when the paper was published. When he returned to Harvard on a permanent basis in the late 1940s, he had already moved on to other areas of mathematics. Extending the technique to higher dimensions has also proved challenging because the approach is so closely linked to the Euler characteristic (X), and the simple Euler formulas for two-dimensional surfaces—such as $X = 2 - 2g$ and $X = V - E + F$ for polyhedrons, where V is the number of vertices, E is the number of edges, and F is the number of faces—do not apply in higher dimensions. Nevertheless, interest in Ahlfors's ideas on covering surfaces has picked up considerably since 1985 or so, according to Drasin.[30] His landmark paper on the subject, adds Osserman, still stands as "one of his masterpieces."[31]

It would be wrong, however, to assume that Ahlfors's inspired work on covering surfaces went unnoticed at the time simply because it did not spur much immediate follow-up activity. To the contrary, his 1935 paper was cited as the principal reason for his being named corecipient of the Fields Medal, a newly created prize to honor "outstanding achievements in mathematics," which was given out at the 1936 International Congress of Mathematicians held in Oslo, Norway. The prize came as a complete shock to Ahlfors, who did not know he was to receive the medal until a few hours before the award ceremony. Ahlfors later said that had he not been planning to attend the conference, George David Birkhoff—the dominant figure in Harvard mathematics at the time, who knew about the upcoming award—"probably would have done something about it," meaning that Birkhoff would have used his considerable influence to make sure that Ahlfors got there.[32]

As mentioned before, Ahlfors published another paper on Nevanlinna theory in 1935. Commenting on this paper, Carathéodory found it surprising that Ahlfors was able to condense the whole of his mentor's theory into just fourteen pages. Two years later, Ahlfors published a paper that brought the Gauss-Bonnet theorem into Nevanlinna theory. The theorem, which was formulated in the nineteenth century by Carl Friedrich Gauss and Pierre Bonnet, connects the geometry (or curvature) of a surface with its topology as described by the Euler characteristic. Osserman regarded Ahlfors's paper on this subject as "probably at least as influential" as his work on covering surfaces, since it paved the way for higher-dimensional versions of the Gauss-Bonnet theory and Nevanlinna theory, which were pioneered by Shiing-shen Chern and Raoul Bott in the decades to come.[33] (Bott is one of the mathematicians featured in Chapter 7.)

In 1938, the University of Helsinki offered Ahlfors a full professorship when a Swedish-language mathematics chair was created. He was grateful for his stay at Harvard, which he said allowed him to "broaden my mathematical knowledge," but he missed his native Finland. Lindelöf, who was then retired, had also urged Ahlfors to come home out of "patriotic duty,"[34] saying that "Finland can export goods but not intelligence."[35]

War broke out a year later, and the situation in Finland—as with many parts of Europe—steadily deteriorated. Helsinki was bombarded, and the university was shut down due to the lack of male students. Ahlfors's family moved in with relatives in Sweden, which remained neutral throughout the war. Ahlfors, meanwhile, who was unable to join the military for medical reasons, wrote what he considered to be one of his best papers, "The Theory of Meromorphic Curves,"[36] from the inside of an air raid shelter.[37] Ahlfors claimed that had it not been for the long hours of confinement in air raid shelters during the so-called Winter War of 1939–40, he would not have had time for the arduous calculations.[38] Ahlfors was "disappointed that years went by without signs that [this work] had caught on," but the paper ultimately did take hold, inspiring important research related to hyperplanes, which are generalizations of planes in higher dimensions.[39]

Conditions on the home front improved for a short while, until 1941, when Germany attacked the Soviet Union, and the Soviet Union responded by attacking Finland, Germany's ally. The Finnish-Russian War continued, off and on, for the next three years, and when Finland signed an armistice with the Soviet Union in 1944, Finland then became an enemy of Germany. Ahlfors concluded that his research career would suffer if he stayed in the country much longer. "Everyone realized that it would be a long time before conditions in Finland would again be normal so that one could do serious research," Ahlfors said.[40]

In 1944, Ahlfors accepted an offered professorship at the University of Zurich, but the hardest part of that deal was getting to Switzerland with World War II still under way. Although Finland was, by then, no longer fighting with the Soviet Union, it was at war with both the Allied forces and Germany. His first step was going to Sweden, and he was only allowed to take ten crowns out of the country. In the hopes of getting more money to pay for their upcoming trip, Ahlfors smuggled his Fields Medal out of Finland and pawned it in Sweden. "I'm sure it is the only Fields Medal that has been in a pawn shop," Ahlfors said.[41] He later recovered the medal with the help of Swedish friends, joking that this might have been about the only practical use for a Fields Medal that he had heard of.[42]

While waiting in Sweden for the opportunity to fly to Scotland, Beurling and Ahlfors, who were now good friends, worked together at the Uppsala University, where Beurling was based. For safety reasons, the Swedes ran stratospheric flights on moonless nights between Sweden and Scotland, but the waits for such flights were long, and one generally needed diplomatic connections to be eligible. Finally, in March 1945, Ahlfors and his family were told to be ready to leave that night, weather permitting. It was a perilous flight, in a nonpressurized aircraft, but they made it to Prestwick, Scotland. They went to London by train and—after about ten tries—succeeded in securing a boat to France. From there they took a train to Switzerland that was bombed along the way by the Germans.

It was an arduous, as well as dangerous, journey to get to Zurich. Unfortunately, upon arriving there Ahlfors was never really happy. "We landed in a never-never land that seemed to have had no contact with the war," he reported. "My first impression was that the university, and perhaps the whole country, had been asleep for a hundred years."[43] Switzerland had been spared from the ravages of war, but rather than striking him as a paradise, it instead felt to him like the country was "in a state of suspension." Ahlfors did not find Switzerland to be an inviting place for strangers, and, apart from the conviviality of some colleagues at the university, he did not feel welcome there.[44]

When Harvard offered him a full professorship, starting in 1946, Ahlfors leapt at the offer, staying on until his retirement in 1977. He left Zurich with a good conscience, knowing that his position at the university had been ably filled by Nevanlinna. He later described his relationship with Harvard—with both the administration and his colleagues in the department—as "singularly happy," saying that the university provided him with "the optimal milieu for my research."[45] Harvard, of course, benefited from Ahlfors's presence, too, as his return to Cambridge helped restore the department's leadership position in classical and complex analysis.

Ahlfors was, in the words of his colleague Bott, "a great Harvard patriot [who] on many occasions . . . delivered himself of the opinion that Harvard was the best of all universities."[46] That said, Ahlfors had absolutely no interest in taking on administrative chores for his beloved university. In fact, when it was his turn to become department chair, he offered to resign instead.[47]

Fortunately, he was able to keep his focus on his research career, which thrived during his Harvard years. Early in his term there, he persuaded his friend Beurling, who was then in Uppsala, to join him at Harvard, which Beurling did in 1948 and 1949 before moving to the Institute of Advanced Study in Princeton. For years, Beurling had been working independently on the notion of the "extremal length" of a set of curves. Ahlfors teamed up with him briefly in Uppsala during World War II and then again at Harvard, helping to clarify, as well as further develop, the concept. Their collaboration culminated in a widely cited paper on this topic published in 1950.[48]

Extremal length, a generalization of the area-length method, applies not to a single curve but rather to a whole collection of curves. Consider, for example, an annulus, the region in a plane bounded between two concentric circles. One could draw all kinds of closed curves inside the annulus that go around the central hole, though for the purposes of this discussion, we will limit ourselves to curves of finite lengths—curves that are not infinitely wiggly. Determining the extremal length for this assortment of curves is a problem in differential geometry. Each curve, of course, has a length, L, as well as an area, A, that it bounds. The calculation, which is based on calculus, involves computing the ratio L^2/A for all possible curves—in all possible "conformal metrics" (or geometries for which angles are preserved during transformations)—and then picking the minimum.

One can also compute the extremal length in more general situations, such as a domain of the so-called annulus type, consisting of the region bounded between two closed curves that do not have to lie in the same plane. In working out the extremal length, mathematicians can exploit the fact that such a domain is conformal to a genuine annulus (bounded by two concentric circles in a plane). They can then use the ratio of the radii of the concentric circles in the regular, conformal annulus to calculate the extremal length of this more general kind of domain.

But why is this concept considered useful? Well, suppose you have a mapping, f, from a domain, D, to another domain, D', each of them being subsets of the plane of the annulus type. D is the input in this case, while D' is the output. Or, framing this in mathematical language, $f: D \to D'$. Let us further suppose that f is a conformal mapping, which means that the angles between curves do not change as you move between the two domains (though the size of objects can change). What this means is that if you take two curves in D, which have a certain angle between them, and apply the mapping to them, you will end up with two curves in D' that might look somewhat different, yet the angle between those curves stays the same. The exciting thing here is that for a conformal mapping, the extremal length of D will always equal the extremal length of its image, D'.

Conversely, suppose that you have two families of closed curves, confined within an annulus, which cannot be shrunk down to a point. (These curves, as stated before, must go around the hole in the center of the disk.) Other than that, you may not know a thing about these curves. But if the extremal length for each family of curves is the same, then you know that the two domains—the two annuli—are conformally equivalent, meaning there is a conformal mapping between the domains that the families belong to. The "flip side" of this statement is also true: if the two annuli are conformally equivalent, then the extremal lengths of the closed curves they bound must be the same as well.

Extremal length has become an important tool for studying problems in complex analysis. Ahlfors himself drew on the technique extensively while developing "quasiconformal mapping"—an idea that occupied much of his attention in the 1950s. But before discussing that concept, we should first mention another noteworthy achievement during that period, the 1953 publication of Ahlfors's book *Complex Analysis,* which he dedicated to his first teacher, Lindelöf. *Complex Analysis* stood as the definitive text in the field for many decades—a book that is still widely used today, more than a half century later. "There are few other texts in modern mathematics that have played such a dominant role," notes Steven G. Krantz, a mathematician at Washington University in St. Louis.[49] "I use his textbook every other year when I teach our graduate course [in complex analysis]," adds Albert Marden. "I think it is still the best available text."[50]

But, as just mentioned, Ahlfors's research interests had shifted into a new direction: quasiconformal mappings, which, as the name implies, are mappings that are somewhat less exacting and restrictive than conformal mappings. Whereas a conformal map, for example, would automatically map an infinitesimal circle to an infinitesimal circle, a quasiconformal map allows for some distortion so that the image of the infinitesimal circle might instead be an infinitesimal ellipse. An ellipse, by definition, has a major axis of length a—the longest line segment between two points on the curve that runs through the center—and a perpendicular axis of length b, which represents the shortest line segment running through the center. The ratio a/b is always greater than one unless the ellipse is a circle—a very special kind of ellipse in which case the ratio is equal to one. "If you don't require the ratio to be one but instead set an arbitrary upper bound—such as one hundred or one million—you'll find there are many more functions that satisfy that condition," Hinkkanen explains. "Having a uniform upper bound can be useful because it gives a mathematician some control over what the function can do."[51]

Things can get even more complicated if the image of the circle is a loop that does not have to be smooth, meaning it does not have to be differentiable at every point, though some constraints are placed on its roughness. Quasiconformal mapping imposes bounds on the degree to which the circle can be distorted. In this case, we are talking about the "local distortion."

Here is where extremal length can come into play: Suppose the input or domain for your function is a set of curves (including circles) that has an extremal length. The class of objects in the image will have an extremal length, too, but that length—because the mapping is no longer conformal—is likely to be different. You can take the ratio of the larger extremal length over the smaller extremal length and set a limit on how big that ratio can be—a limit that applies to all the objects in the domain and image—thereby establishing a "global constraint" on how much distortion is permitted.

Ahlfors coined the term "quasiconformal mappings" in his celebrated 1935 paper on covering surfaces, commenting decades later: "Little did I know at the time what an important role quasiconformal mappings would come to play in my own work."[52] Although Ahlfors was the first to employ the term, the German mathematician Herbert Grötzsch had introduced the concept seven years earlier, while proving some interesting properties as well. The Soviet mathematician Mikhail Lavrentev—whose work was more closely connected with partial differential equations than with function theory—independently, and for different reasons, came up with essentially the same class of functions in 1935. "Yet Ahlfors was the first to show that quasiconformal mappings provide an efficient tool in complex analysis," says Lehto, "and he gave an indication of their intrinsic role in the theory of analytic functions and Riemann surfaces."[53]

Ahlfors's 1954 paper, "On Quasiconformal Mappings," notes Cornell mathematician Clifford J. Earle, a former Ahlfors Ph.D. student at Harvard, "brought international attention to the subject and really opened up the field."[54] "Ahlfors's work here set off a chain reaction in mathematics that is still continuing," adds Marden.[55] Lehto agrees with that assessment: "Among Ahlfors's papers, this one, if any, was seminal in leading to much new research."[56]

Quasiconformal mapping, according to University of Michigan mathematician Frederick Gehring, offers "a stripped-down picture of the geometric essentials of complex function theory and, as such, admits applications of these ideas to many other parts of analysis and geometry. They constitute just one illustration of the profound and lasting effect that the deep, central, and seminal character of Lars Ahlfors's research has had on the face of modern mathematics."[57]

Ahlfors and Beurling collaborated on a paper about quasiconformal mappings that was published in 1956—the aftermath of which led to a serious rift in their relationship.[58] While scrupulously giving due credit to Beurling, Ahlfors nevertheless did something that irritated his colleague—either by speaking about the paper before Beurling wanted him to or presenting the ideas in a manner that Beurling (who had the reputation of being extremely particular) did not like. The upshot of it was a hiatus in their friendship, lasting a couple of decades, during which time neither man spoke to the other or communicated in any way. Ahlfors tried to explain how their longstanding friendship got derailed in an article he wrote after Beurling's death in 1986. After Beurling left Harvard for his job in Princeton, Ahlfors wrote, "With the loss of proximity, too many frictions arose that could not be put right at once. Misunderstandings were piled on misunderstandings, and in the end nobody could do anything about it." At a 1985 conference at Purdue, where both had been invited as speakers, Ahlfors and Beurling—who were then about eighty years old—finally reconciled their differences. "Arne's health was known to be brittle, and we were not sure that he would be able to make it," Ahlfors said. "But he came, and the old magic worked. He put his hand on my shoulder, and with no word spoken I knew that bygones were bygones . . . We parted as friends."[59]

Although Ahlfors and Beurling's collaboration on quasiconformal mappings came to an abrupt end in the mid-1950s, Ahlfors continued to pursue the subject with great vigor. He found inspiration in the work of Oswald Teichmüller, which shed new light on Riemann surfaces. "It had become increasingly evident that Teichmüller's ideas would profoundly influence analysis and especially the theory of functions of one complex variable," Ahlfors wrote.[60]

Interestingly, Nevanlinna spent the 1936–37 academic year at the University of Göttingen, where Teichmüller—who was then an assistant in the department—spoke with him and attended his lectures. Teichmüller gained much from these interactions, getting clues as to how quasiconformal mappings might relate to Nevanlinna theory. Nearly two decades later, Ahlfors picked up on those ideas, which by then were mostly forgotten, and carried them far, helping to forge new pathways in complex analysis that persist to this day. Viewed with hindsight, says Drasin, "the historical accident of Nevanlinna being in Germany for a year changed the entire history of complex analysis."[61]

Teichmüller's work was not well known at the time, mainly because he was an ardent Nazi, political fanatic, and Hitler zealot who was killed in 1943 in combat during World War II. Many of Teichmüller's most important papers, moreover, were published in *Deutsche Mathematik*, a journal dedicated to the Nazi cause in which scholarly articles were mixed in with race (or racist) propaganda. While Ahlfors found Teichmüller's political views to be morally repugnant, he was still able to recognize the importance of Teichmüller's mathematics. "Ahlfors hated the man but respected the work, giving credit where credit was due," commented University of Minnesota mathematician Dennis A. Hejhal.[62] Ahlfors then set out to put this body of work—and the tantalizing ideas contained therein—onto a firmer foundation, providing rigorous proofs for many of Teichmüller's most important hypotheses and results. "Teichmüller had written some brilliant papers in the subject, but his proofs were not complete," comments Marden. "Ahlfors recognized their importance and developed the theory from scratch so that it could be understood."[63]

Teichmüller had sketched out the beginnings of a general theory that concerned entire families of Riemann surfaces. "At an age when most mathematicians are past their prime, Ahlfors took up this theory, adding some very important geometric ideas of his own," and, in concert with Lipman Bers, "not only laid down the foundations of his theory but established a flourishing international school of research around it, with Ahlfors remaining the dominant figure until his retirement in 1977," his Harvard colleagues wrote.[64]

The relationship between Ahlfors and Bers started in an interesting way. While working on quasiconformal mappings, Bers made use of an inequality in a paper by Lavrentev that had been attributed to Ahlfors. At around the same time, Ahlfors gave a lecture at Princeton in the course of which he proved a theorem surrounding this same inequality. After the talk, Bers asked him where he had published the paper that established this inequality. Ahlfors confessed that he never had. "Then why did Lavrentev credit you with it?" Bers asked. Ahlfors replied: "He probably thought I must know it and was too lazy to look it up in the literature." When Bers met up with Lavrentev three years later, Lavrentev admitted that Ahlfors's assessment was indeed correct. "I immediately decided," Bers said, "that, first of all, if quasiconformal mappings lead to such powerful and beautiful results and secondly, if it is done in this gentlemanly spirit—where you don't fight over priority—this is something that I should spend the rest of my life studying."[65]

The professional collaboration between Ahlfors and Bers was "spiritually very close," wrote University of Connecticut mathematician William Abikoff. "They were in constant contact," for the most part working independently, and often simultaneously, on similar ideas. They wrote only one paper together, which Bers discussed at the 1958 International Congress of Mathematicians in Edinburgh, Scotland.[66] At that meeting, Bers presented his and Ahlfors's new proof of the celebrated Riemann mapping theorem or "uniformization theorem"—an exercise that relied heavily on quasiconformal mapping.[67] First stated by Riemann (a student of Gauss) in the early 1850s, the mapping theorem held that a simply connected Riemann surface (which is to say, a surface without any holes) is conformally equivalent to either the open unit disk (the set of points inside a unit circle, excluding the circle itself), the complex plane, or the Riemann sphere (which is the complex plane with an additional point at infinity). That a simply connected Riemann surface is conformally equivalent to these other domains means that one can find a conformal map that takes every point on the Riemann surface to every point on the open unit disk, complex plane, or Riemann sphere. The theorem further states that one can find a geometry (or metric) for the Riemann surface that has constant curvature, just as a simple sphere has constant (positive) curvature.

Bers was the first to acknowledge that they had overlooked the prior, fundamental work of (former George D. Birkhoff student) Charles Morrey and failed to accord him proper credit in their joint 1960 paper. "The great difference in language and emphasis had obscured the relevance of Morrey's [1938] paper for the theory of quasiconformal mappings," Ahlfors said many years later by way of explaining their oversight.[68] (The mathematician John Milnor has subsequently called the theorem the Morrey-Ahlfors-Bers measurable mapping theorem to make sure that Morrey was given due credit for proving this theorem, and publishing that proof, more than twenty years before Ahlfors and Bers.)[69]

One of the things Ahlfors was focusing on at the time, which was evident in his work with Bers on the uniformization (or mapping) theorem, was using quasiconformal maps to study deformations of Riemann surfaces. As discussed earlier, a quasiconformal mapping from a compact Riemann surface to a deformed Riemann surface can be distorted to a certain extent. Because Riemann surfaces, in turn, can be "represented" by so-called Kleinian groups—an idea that will be elaborated upon shortly—Ahlfors and Bers, among others, began investigating these groups as part of a fruitful research enterprise that is still being carried on by their "descendants" today.

The theory of Kleinian groups was originally developed by Felix Klein and Henri Poincaré, with Poincaré coining the term, said Ahlfors, "much to the displeasure of Klein."[70] (Klein, as described in Chapter 2, was a mathematician based at the University of Göttingen who, in the late 1800s, trained the Harvard scholars Maxime Bôcher, Frank Nelson Cole, and, for a time, William Fogg Osgood. Poincaré is a legendary figure in mathematics whose name comes up frequently throughout this book.)

A Kleinian group is a group of two-dimensional transformations of the "extended complex plane." To break down that rather ponderous statement, we will start with the extended complex plane, which is the complex plane with an additional point lying at infinity. The extended plane can be thought of as a sphere—technically a "Riemann sphere"—with the north pole representing the point at infinity, as in our prior discussion of covering surfaces. The transformations of the Kleinian group that act on this plane or sphere—a subgroup of so-called Möbius transformations—are functions of a complex variable, z, of the form $f(z) = (az + b)/(cz + d)$, where a, b, c, and d can be either real or complex numbers. A 2×2 matrix can represent this function:

$$\begin{bmatrix} a & b \\ c & d \end{bmatrix}$$

subject to the restriction that the "determinant" $(ad - bc)$ is not equal to zero. A further restriction is imposed by the fact that a Kleinian group is a "discrete" group, meaning that a, b, c, and d cannot vary continuously and can instead assume only certain values.

That is all rather abstract, of course, so what exactly do these transformations do? They include, for instance, translations, such as shifting the origin on the plane by one unit to the left or right, while infinity stays fixed; rotations of a certain angle; or a magnification by 2, so that 2 becomes 4, 4 becomes 8, and so forth. All of these actions—translations, rotations, and magnifications—are possible, and they can be combined in practically all manner of ways.

To understand how a Kleinian group "represents" a Riemann surface, explains Stony Brook University mathematician Irwin Kra, a former student of Bers's, "you first have to see how a group acts on a space. When a Kleinian group acts on a space (such as a plane or a sphere), it divides the space into two regions: one, called the 'limit set,' where the space behaves badly, and two, a so-called 'region of discontinuity,' where it behaves in a better, more controlled manner."[71]

Suppose, for instance, that the region of discontinuity is a rectangle, which happens to be a very simple Riemann surface. If you glue together two opposite sides of this rectangle—an operation that is considered "legal" in this game—you get a cylinder, which is another, slightly more complicated Riemann surface. The ends of this cylinder can be glued together, in turn, to make a doughnut or torus, which is another Riemann surface. Alternatively, you might start with an octagon, rather than a rectangle, and through a similar gluing process you might end up with a "double pretzel," which is a Riemann surface of genus 2, meaning that it has two holes.

"The elements of the [Kleinian] group give you instructions for selecting a subset from the well-behaved set [the region of discontinuity], and they also give you instructions on how you can glue the sides together," Kra continues. "If you go through this procedure, it leads you to a certain set of objects, and those objects are Riemann surfaces."[72]

This is precisely where Ahlfors made a great breakthrough. First you have to start with a finitely generated group that acts discontinuously on some regions of the sphere. "Discontinuous" in this case means that if you take a point in the domain D and subject it to a group of transformations, you will get a set of points in the image domain, D', and those points cannot be too densely packed. "Finitely generated" does not imply that the group itself has a finite number of elements (because the groups Ahlfors was interested in were not "finite" in that sense). Rather, it means that the group is generated from a finite number of elements. For example, all of the integers can be generated from a single element, the number 1. If you start with 1 and add 1 to it, and then add 1 to that, and so on, you can eventually get all of the counting numbers or positive integers. Similarly, if you start with 1 and keep subtracting by 1, you will get the rest of the integers. Satisfying those two conditions—that the group is finitely generated and acts discontinuously on some regions of the plane or sphere—ensures that the group is Kleinian. Ahlfors then proved that if you follow the procedure discussed above—constructing Riemann surfaces from the parts of the sphere where things behave nicely, with some allowed cutting and gluing—you will get only a finite number of Riemann surfaces. A Kleinian group, in other words, represents only a finite number of Riemann surfaces.

This finding, the "Ahlfors finiteness theorem," which appeared in a 1964 paper, constituted a major advance.[73] At the start of the twentieth century, Poincaré had suggested a program for studying Kleinian groups, which had not yet born fruit. "In the early sixties, not much was known about Kleinian groups," said Kra. "Research in the field seemed to be stuck and going nowhere. Ahlfors completely ignored Poincaré's program and took a different route to prove the finiteness theorem."[74]

"Poincaré thought you needed three-dimensional techniques, but Ahlfors showed that there are many problems concerning Kleinian groups that can be solved with two-dimensional techniques," Kra added. "That was Ahlfors's genius. The course charted by Poincaré was too difficult for people to make progress with until much later."[75] By abandoning Poincaré's original program, which struck many as a great departure from conventional wisdom, says Kra, Ahlfors was able to achieve "the most significant result in this subject in over fifty years."[76] The subject, adds Marden, "which began with Poincaré, was awoken from its long somnolence by Ahlfors's brilliant discovery."[77]

Important as that accomplishment may have been, Ahlfors admitted in his 1964 paper that "perhaps of greater interest are the theorems I have not been able to prove."[78] Foremost among these was the so-called measure zero conjecture, spelled out in that same paper, in which Ahlfors theorized that the limit set of a finitely generated Kleinian group—the poorly behaved subset of the plane or sphere—must have zero area in two dimensions. Ahlfors made some progress on this problem but was unable to prove it himself, though his conjecture was subsequently proved (using the theory of so-called hyperbolic three-manifolds).

Ahlfors first discussed the measure zero problem at a 1965 conference at Tulane University. This was the first of what eventually became periodic meetings, called Ahlfors-Bers Colloquia, that are held to the present day, more than forty-five years later, for researchers in fields related to the mathematical interests of Ahlfors and Bers. These colloquia have been held every three years since 1998, constituting an important part of the legacy of these two mathematicians, who struck up such a remarkable, and enduring, collaboration. Indeed, over the years, Ahlfors and Bers had become so closely identified in some people's eyes, notes Kra, that "a group of young Russian mathematicians grew up thinking that 'Ahlfors-Bers' was one person. Many times after arriving in the United States, the first thing they wanted to do was to meet not Lipman Bers, not Lars Ahlfors, but Ahlfors-Bers. It was difficult to comply with this request."[79]

After the mid-1960s, Ahlfors's interests gradually shifted to higher-dimensional problems involving quasiconformal mappings and Möbius transformations. These questions, wrote Lehto, "were to dominate the last period of his research."[80]

Ahlfors officially retired from Harvard in 1977 but stayed active in mathematics for the rest of his life. Liberated from the pressure to "publish or perish" during his retirement, he devoted himself to keeping up with the latest developments in his field. At the age of seventy-five, for example, he sent a proposal to the National Science Foundation that was remarkably concise, consisting of just a single sentence: "I will continue to study the work of Thurston."[81] (William Thurston, a mathematician based at Cornell University until his death in 2012, showed how Kleinian groups could be used to study three-dimensional manifolds—a celebrated result that drew on Ahlfors's finiteness theorem.)[82]

Interviewed when he was eighty-four, Ahlfors discussed the challenges of trying to do original work in mathematics at an advanced age. "I know of mathematicians who no longer do mathematics because they are afraid that it would not compare with what they have done before. I am not afraid of that. I can see that I have a hard time, and I can see that I make mistakes. But I always find the mistakes and learn from them . . . I am still doing something that I think will be good—still confident that it will be good mathematics when it comes out." As always, Ahlfors relied heavily on the subconscious while doing mathematics, admitting that he sometimes had success working in bed, especially first thing in the morning when he was able to think much more clearly. "One works, and works hard, but one does not really discover anything while working. It is later that one makes discoveries."[83]

Although he was a notoriously reserved man, Ahlfors and his wife, Erna, were famous for the parties they threw. Known for consuming appreciable quantities of alcohol, Ahlfors tended to become increasingly less reserved as the night wore on. But he always insisted on maintaining some level of decorum at his parties, and one of the rules that his guests had to abide by was not talking about mathematics—or engaging in other forms of shoptalk—during an Ahlfors social event.

"Lars was also a man of action and had an instinct for achieving his ends in the simplest way, be it in the exposition of his mathematics or in real life," his Harvard colleague Raoul Bott observed. When Ahlfors was accosted by a knife-wielding assailant at the entryway to his apartment on Boston's Beacon Hill, Bott said, "He did not hesitate. It happened that he was returning from a shopping expedition with a bottle of whiskey (first-class, of course) under his arm. So he instinctively hit the chap over the head with the bottle and managed to open his door to safety before the assailant recovered his wits. Thereafter, it was mainly the loss of that fine whiskey that Lars lamented."[84]

Lehto recounted an episode that took place at a mathematics conference in Romania, which he felt might have offered a revealing glimpse into Ahlfors's personality. During a roundtable discussion of philosophical matters—the kind of discussion Ahlfors normally avoided—he was asked about the most important thing in life. "There is no doubt about that: It is alcohol," he said with a straight face, bringing all conversation in the room to an abrupt halt. When asked what was the second most important thing, he said that question was more difficult to answer, suggesting that second place would be shared by mathematics and sex.[85]

Although mathematics may have placed second on Ahlfors's list of priorities, that is certainly what he will be remembered for, as he initiated numerous avenues of mathematical research that are still thriving. In winning the 1981 Wolf Prize for "outstanding" accomplishments in mathematics, which he shared with his Harvard colleague Oscar Zariski, Ahlfors was credited with combining "deep geometric insight with subtle analytic skill," the Wolf Foundation judges found. "Time and again he attacked and solved the central problem in a discipline. Time and again other mathematicians were inspired by work he did many years earlier. Every complex analyst working today is, in some sense, his pupil."[86]

After his death in 1996, Ahlfors received accolades from colleagues from all over the world. The memorial service held for him reflected the personality of the man being honored. "Ahlfors was exceedingly reserved and reticent about his work," Hejhal said. "He was also very serious and very careful, always adhering to extremely high standards. In contrast to other memorial services, where people get up and tell funny stories about the deceased, people spoke carefully and respectfully about Ahlfors because that's the kind of man he was."[87]

In a posthumous tribute published in *The Notices of the American Mathematical Society*, Ahlfors was called "arguably the preeminent complex function theorist of the twentieth century."[88] In a similar vein, his Harvard colleagues noted that "the global, geometric view of complex variable theory, which Ahlfors had championed so brilliantly throughout his career, remains a center of research both in pure mathematics and in the string theory of physics."[89]

Ahlfors will be remembered as a devoted teacher with a booming, deep bass voice. "Nobody ever fell asleep in an Ahlfors lecture," said Hejhal. "It was impossible because he spoke so loudly. You often wanted to say, 'Please dial it down,' but nobody ever did."[90]

On a more serious note, Hejhal added, he always considered Ahlfors as a person who truly believed in excellence: "Someone who not only would not lower the bar but who would instead raise it to facilitate bringing out talent that he intuited was there." Hejhal began corresponding with Ahlfors while he was a high school student and later, briefly, became Ahlfors's colleague when he took a two-year Benjamin Peirce Assistant Professorship at Harvard. He felt extremely fortunate to have had such an encounter during his formative years. "One could not have asked for a better start," Hejhal said.[91]

Stanford mathematician Robert Osserman also got his start in the field with Ahlfors as his graduate adviser. Osserman had been told to seek out Ahlfors by Vidar Wolontis, a fellow Harvard student of Finnish descent and an Ahlfors protégé, who urged him to "look at Ahlfors's bibliography . . . Only a relatively small number of papers, but every one significant." Osserman came to believe that Ahlfors's unspoken motto—like the adage attributed to Gauss of "few, but ripe"—was "few, but outstanding." That approach, Osserman said, "appealed to me greatly, and I did become Ahlfors's student—a decision I have always felt lucky to have made."[92]

Ahlfors, said Marden, "made substantial contributions to so many topics that pretty much everyone was touched by his work. For many years, Lars was the center of the solar system of complex analysts, which was held together by his gravity."[93]

Longtime friend Lehto remarked that Ahlfors was an active researcher for an exceptional length of time. More than just a leading name in complex analysis, having made fundamental contributions to the field over the course of fifty years, "he was time after time a pioneer, a prophet." For finding out what developments in modern complex analysis were worth pursuing, Walter Hayman, a mathematician at Imperial College London, offered a simple suggestion: "Watch Ahlfors."[94]

FIGURE 1 Benjamin Peirce
(Courtesy of Harvard University Archives)

FIGURE 2 Charles Sanders Peirce
(Courtesy of Harvard University Archives)

FIGURE 3 William Fogg Osgood
(Courtesy of Harvard University Archives)

FIGURE 4 Maxime Bôcher
(Courtesy of Harvard University Archives)

FIGURE 5 George David Birkhoff
(Courtesy of Harvard News Office)

FIGURE 6 Garrett Birkhoff
(Courtesy of Harvard University Archives)

FIGURE 7 Marston Morse
(Photo by Ulli Steltzer. From The Shelby White and Leon Levy Archives Center, Institute for Advanced Study, Princeton, N.J., USA.)

FIGURE 8 Hassler Whitney
(Photo by Herman Landshoff. From The Shelby White and Leon Levy Archives Center, Institute for Advanced Study, Princeton, N.J., USA.)

FIGURE 9 Saunders Mac Lane
(Copyright University of Chicago)

FIGURE 10 Lars Ahlfors
(Photo by David J. Lewis, Harvard
Yearbook Publications, Inc.)

FIGURE 11 Andrew Gleason
(Courtesy of Harvard News Office)

FIGURE 12 George Mackey
(Courtesy of Harvard News Office)

FIGURE 13 Oscar Zariski
(Photo by David L. Crofoot, Harvard
Yearbook Publications, Inc.)

FIGURE 14 Shreeram Abhyankar
(Photo courtesy of Yvonne M.
Abhyankar)

FIGURE 15 Heisuke Hironaka
(Photo by Martha Stewart)

FIGURE 16 David Mumford
(Photo by Peter Norvig)

FIGURE 17 Michael Artin
(Photo © Carolyn Artin. All rights reserved.)

FIGURE 18 Richard Brauer
(Photo courtesy of the Bentley Historical Library, University of Michigan)

FIGURE 19 Raoul Bott
(Courtesy of Harvard News Office)

FIGURE 20 John Tate
(Photo by Vincent B. Wickwar, Harvard Yearbook Publications, Inc.)

FIGURE 21 Barry Mazur
(Photo by Jim Harrison)

6

THE WAR AND ITS AFTERMATH: ANDREW GLEASON, GEORGE MACKEY, AND AN ASSIGNATION IN HILBERT SPACE

The bombing of Pearl Harbor brought World War II home to the United States in a dramatic, as well as deadly, fashion. Following the attack on December 7, 1941, which left 2,400 Americans dead and thirty U.S. ships damaged or destroyed, the country immediately went to war, issuing a formal declaration the next day. More than ten million American men and women enlisted in the military during this struggle—a massive involvement that extended deeply into academia. Harvard was no exception. Its math department thinned out significantly during World War II, with mathematicians joining the armed forces or volunteering as researchers in support of the Allied cause. Indeed, aiding the war effort became a top priority, if not *the* top priority, at Harvard, as at many universities throughout the country.

Shortly after Pearl Harbor, Harvard professor Marshall Stone became chairman of the American Mathematical Society's newly formed War Policy Committee. John H. Van Vleck, who had joint appointments in Harvard's physics and math departments, led a group at the school's Radio Research Laboratory, developing radar countermeasures. Van Vleck also participated in the Manhattan Project, winning a Nobel Prize in Physics several decades later. Joseph Walsh, who first joined Harvard's math faculty in 1917, reenlisted in the U.S. Navy from 1942 to 1946—serving as a lieutenant commander and then a commander—after having been in the Navy twenty-five years earlier during World War I. Julian Coolidge, who became an instructor at Harvard in 1899 and a professor a few years later, came out of retirement—at the age of almost seventy—to teach calculus at Harvard in order to cover for fellow faculty members who were engaged in the defense of their nation.

Saunders Mac Lane, as discussed in Chapter 4, headed the Applied Mathematics Group, based at Columbia University, which worked on numerous war-related problems. Members of this group included Harvard topologist Hassler Whitney; Irving Kaplansky, Mac Lane's former Ph.D. student who was a Benjamin Peirce Instructor at Harvard at the time; and George Mackey (about whom much more is said in this chapter), a Harvard instructor who had recently gotten his Ph.D. under Stone's tutelage and was soon to become a permanent faculty member.

Mac Lane described one of the projects the Applied Mathematics Group took on for the U.S. Air Force—a case where the principles of mathematics led to a crucial, though counterintuitive, result. The question they addressed concerns how a gunner on a U.S. bomber plane should aim at an attacking fighter plane. "The fighter is approaching the bomber, but the bomber is moving forward at the same time," he explained. "The resulting rule is that the gunner should aim toward the tail, which is opposite of the rule for hunting ducks, where the rule for a stationary hunter is to aim ahead of the ducks. A major part of our problem was properly training machine gunners to aim toward the tail."[1]

According to a story that Vassar mathematician John McCleary heard from multiple sources, a gunner on an American plane, who was engaged in a dogfight with two enemy planes, "fired in front of the second plane, as intuition would dictate, but hit the first plane." If true—and McCleary was unable to substantiate this account—this anecdote would certainly bolster Mac Lane's statement about the need to retrain gunners in the U.S. military on the basis of the teachings of geometry.[2]

In research sponsored by the Navy's Bureau of Ordnance, Garrett Birkhoff worked with his Harvard math colleague Lynn Loomis and MIT mathematician Norman Levinson, trying to predict the underwater trajectories of air-launched torpedoes—a problem in which his father, George David Birkhoff, also took an interest.[3] The younger Birkhoff also joined a committee, along with Marston Morse and John von Neumann, that was charged with analyzing ways of improving the effectiveness of antiaircraft shells. As a consultant to the Ballistics Research Laboratory at the Aberdeen Proving Ground in Maryland, Garrett Birkhoff studied issues involved in penetrating tank armor.

As a result of this research, he became increasingly interested in applied mathematics. "It was clear to me that our war effort was unlikely to be helped by any of the beautiful ideas about 'modern' algebra, topology, and functional analysis"—or his long-standing devotion to lattices and groups—"that had fascinated me since 1932," so Birkhoff began to focus on "more relevant topics."[4] And even after the war had ended, he continued to work on problems of naval research. Unlike many of his colleagues, who promptly resumed their full-time study of abstract mathematics at the war's conclusion, Birkhoff pursued a mixture of pure and applied math, sensing that "the ivory tower of prewar academe was not likely to return for many years, if ever during my lifetime."[5]

Stanislaw Ulam also made the shift during the war away from pure mathematics and toward the applied. At the suggestion of George D. Birkhoff, the Polish-born Ulam spent 1936–1940 at Harvard in the Society of Fellows and as an instructor in the math department. Birkhoff tried to get Ulam a permanent appointment, but his Harvard colleagues did not support Ulam's candidacy, perhaps owing to his short publications list at the time. So instead, Birkhoff helped land him a teaching job at the University of Wisconsin.[6] Soon afterward, Ulam was recruited for the Manhattan Project, where he and von Neumann—among the few mathematicians working on the project—teamed up to perform elaborate numerical calculations that helped lead to the successful design of the first atomic bomb.

Anticipating the invaluable chores that computational machines would eventually be able to perform, Birkhoff helped secure funding for the Harvard physicist and computer scientist Howard Aiken to develop the world's largest and most powerful calculator, the Harvard Mark I. Built and housed at Harvard, this programmable device, eight feet tall and more than fifty feet long, was unveiled in 1944 and subsequently used for gunnery and ballistics calculations, as well as for calculations in the Manhattan Project.[7] Portions of the Mark I, which is considered the world's first mainframe computer, remain on display in Harvard's Science Center, just below the school's math department.

Ulam later invented the Monte Carlo method for solving mathematical problems by statistical means. He also made critical contributions (no pun intended) to the development of the hydrogen bomb, working through calculations by hand, which showed that a method of construction proposed in 1949—and favored at the time—would not work. This result was later confirmed by ENIAC, the fastest computer then available, which was developed, in part, to aid the design of thermonuclear weapons and work through the requisite hydrodynamical calculations. Referring to the calculations Ulam carried out with paper and pencil, the physicist Edward Teller quipped that "in a real emergency, the mathematician still wins—if he is really good." Ulam later solved the problem of initiating fusion in the hydrogen bomb, demonstrating that a differently configured weapon would work. Once again, his calculations beat what was then the fastest computer around, SEAC.[8] Although he had outdueled the best machines on at least two occasions, Ulam, along with von Neumann, still argued for the creation of an advanced computational facility at Los Alamos—a center that was formed shortly thereafter and exists to this day. Ulam continued to work in Los Alamos's theoretical division, interspersed with some stints in academia, until 1967.

The creation of the atomic bomb—for better or worse, and regardless of its role in bringing an end to the war—was mostly an exercise in physics, carried out by physicists, with mathematicians playing an important though ancillary part. "The most sensational achievements of mathematics during the War were probably in ciphers and codebreaking," wrote J. Barkley Rosser, former director of the Army Mathematics Research Center at the University of Wisconsin–Madison.[9] A young American mathematician named Andrew Gleason had a big impact in this area.

When the Japanese struck Pearl Harbor, Gleason had not yet made his way to Harvard, where he would ultimately remain for nearly fifty years and spend his entire academic career. He was still an undergraduate at Yale who had decided to go into mathematics at an early age—just as soon as he realized he was not destined to be a fireman.[10] Displaying an extraordinary mathematical talent in college, Gleason fared well in graduate-level courses and placed in the top five in the nationwide William Lowell Putnam Mathematical Competition for three successive years—1940, 1941, and 1942—a feat rarely accomplished in the competition's seventy-five-year history.

He had a gift for solving problems and an uncanny ability to carry out elaborate calculations in his head. Armed with this knack for "quick and dirty mathematics," as he put it, which was "something that a lot of pure mathematicians don't know how to do," Gleason was able "to make a quick appraisal as to whether there is a sufficient statistical strength in a situation so that hopefully you will be able to get an answer out of it."[11] In other words, he was just the sort of person the U.S. Navy's cryptanalysis group in Washington, D.C., was looking for. He joined the team in June 1942, immediately upon graduating from Yale, quickly establishing himself as a leader in that effort, despite his youth—a twenty-year-old when he started out, playing in a contest where the stakes and pressure could not have been higher.

The U.S. naval communications group known as OP-20-G worked in concert with the British code-breaking group located in Bletchley Park, whose most renowned member was Alan Turing, a legendary mathematician credited with inventing the modern computer and the field of artificial intelligence, among other achievements. The British team led the operation, as they had been actively working on deciphering German codes for years before the United States entered the war. One problem the Americans took on, called Seahorse, involved communications sent between Germany's naval headquarters in Berlin and its attaché in Tokyo.

To disguise their messages, the Germans used a so-called Enigma encryption machine, which relied on alphabetic substitution—switching one letter for another—but it did so in a sophisticated way. Each time a letter was typed into a keyboard, the substitution pattern was changed, so that the same letter in the original message could appear as many different letters in the scrambled cipher text. The complexity of the encryption stemmed, in large part, from the use of multiple wheels or rotors—typically three or four—each of which had twenty-six positions that corresponded to letters A through Z. The wheels were lined up in series, like bicycle wheels sharing a common axis, and they could turn individually or en masse. Each time a letter was typed as part of the original message, one or more of the rotors moved a step, which made the substitution patterns extremely complicated. At the end of this scrambling process, the original message looked like a meaningless string of letters—unless the person at the receiving end had the same Enigma machine and knew how the wheels were set and the electronics configured. Each day, moreover, these settings were typically changed.

To decipher the messages intercepted on a given day, the British and American cryptanalysts had to figure out the exact machine setting for that day out of an estimated 10^{20} possible configurations.[12] Turing and his colleagues were just then developing some of the earliest computers—devices called "bombes" that were crude by today's standards but still incorporated some of the logical architecture found in contemporary computing machines. Since the bombes could not possibly check through all the conceivable settings in order to interpret a message, the code breakers had to employ logic, statistics, probability, and other mathematical tools to narrow down the odds into something more manageable.

Their task was simplified by the fact that, owing to the way Enigma was set up, no letter such as "A" could be encrypted as itself. Over time, the analysts also came to recognize common words and phrases that appeared frequently, which enabled them to guess parts of the message and thereby reduce the immense challenge they faced.

"These early computers, or bombes, were faster than people but far slower than today's computers," explains University of Massachusetts–Boston mathematician Ethan Bolker, a former Ph.D. student of Gleason's at Harvard. "Suppose, for the sake of argument, that there were a million possibilities. Well, maybe the machine was only fast enough to try out 1,000, meaning that you had to be clever enough to figure out which 1,000 to do. That's one of the things Andy was really good at."[13]

For example, Gleason deduced and statistically tested the hypothesis that the wheel settings for messages sent from Tokyo to Berlin started with the letters A through M, whereas messages going in the opposite direction started with the letters N through Z. Based on this finding, Gleason and his colleagues derived new equations, which made their computers (or bombes) more efficient, as well as more productive. Combined with other advances made by code breakers on both sides of the Atlantic, the Allies were able to interpret communications between Berlin and Tokyo consistently in 1944 and 1945 and somewhat more sporadically before then.[14]

Following their successes with Seahorse and other Enigma-related challenges, Allied code breakers turned to Japanese naval codes, including one produced by a machine cipher known as CORAL and another book-based code called JN-25. The team had great success in decrypting these codes as well, with Gleason making important mathematical contributions to this effort. (When questioned about that twenty-five years later by Harvard undergraduates, who heard that Gleason had played a role in cracking the Japanese code, he replied in a modest and characteristically understated manner: "It would not be entirely incorrect to say so.")[15]

While visiting the Washington, D.C.–based team in 1942 and briefing them on the latest methods being used in Bletchley Park, Turing was particularly impressed by "the brilliant young Yale graduate mathematician, Andrew Gleason." (The terms "brilliant" and "genius," of course, were universally applied to Turing himself.) Once Gleason took Turing to a restaurant in D.C., where they talked about how one might estimate the total number of taxicabs in a city simply by looking at a random sampling of the cabs' registration or medallion numbers. When a man sitting nearby expressed dismay over their talking in public about sensitive technical matters, Turing asked: "Shall we continue our conversation in German?"[16]

Although their chat about taxicabs might have sounded frivolous, it related to an analogous problem of considerable import: the Allied forces needed to find out how many tanks Germany could produce in a year to better assess whether their attacks along the Western front might ultimately succeed. Some of the best information they had to draw on was the serial numbers of captured German tanks, which were numbered in the order they were produced (although sometimes those numbers were disguised by a code). Based on the serial numbers so obtained, statisticians produced a far more accurate estimate of German tank production capacity than was contained in previous Allied intelligence reports, which had overestimated actual production by a factor of about five.[17] The mathematical approach helped inform the land campaign in Europe, and this is almost certainly what Gleason and Turing had on their minds when they spoke of taxicabs.

All told, the British and American team's success in cracking the German and Japanese codes was a feat of great significance, credited with shortening the war by perhaps as much as two years, thereby saving thousands of lives.[18] It also meant that Allied soldiers could go home earlier, and the same held for the code breakers, too.

Gleason, for example, left the Navy in 1946, four years after entering, to resume his academic career—this time in Cambridge rather than in New Haven—although he returned to active duty in cryptanalysis during the Korean War. He officially retired from the Navy, at the rank of commander, in 1966 but continued to advise the government's intelligence community until around 1990—serving his country for a span of about fifty years.[19] During this time, Gleason introduced an array of vital mathematical techniques for cryptanalysis, mixing his work on coding theory with a broad range of research on "pure" mathematical topics such as analysis, combinatorics, discrete mathematics, graph theory, measure theory, projective geometry, analytic geometry, and algebra. "Gleason was also one of the rare breed of mathematicians who didn't stay on just one side of the Pure mathematics/Applied mathematics 'divide,'" noted Benedict Gross and other Harvard colleagues. "In fact, his work and attitude gave testimony to the tenet that there *is* no essential divide."[20]

Gleason became a junior fellow in Harvard's Society of Fellows in 1946, thanks in large part to a recommendation from Harvard astronomer Donald Howard Menzel, who headed a naval intelligence division during the war. Menzel's section in the Navy was right next door to Gleason's, and the two spoke extensively about the problems with radar that Menzel was then grappling with. Menzel evidently appreciated the young mathematician's input.

The fellowship, as Gleason described it, typically took "bright people who were just beginning—or maybe just a year into—graduate school and thus got them out of the graduate school 'vise,' as they called it." The program, for him personally, offered both benefits and drawbacks: "The good thing was that I could do what I wanted, and that worked out OK. The bad thing was also that I could do what I wanted." He was somewhat isolated from other mathematics graduate students at similar stages in their careers, and, he said, "it was also probably bad that I didn't really have anybody looking down my throat to see whether I really crossed all the i's and dotted all the t's, or whatever it is you do. There are a lot of technical things I might have learned, and probably should have."[21]

While the latter might be true, and perhaps Gleason *could* have learned more, it is also true that he still managed to acquire an amazing grasp of mathematics, which extended to practically all branches of the field. "One thing that impressed me strongly about Andy was that he understood, in detail, every colloquium we attended independently of the subject matter," said Brown University mathematician John Wermer, a former Ph.D. student of George Mackey's.[22]

These sentiments were echoed by Vera Pless, a mathematician at the University of Illinois at Chicago, who fondly recalled the seminars on coding theory that Gleason ran in the 1950s: "These monthly meetings were what I lived for. No matter what questions we asked him on any area of mathematics, Andy knew the answer. The numerical calculations he did in his head were amazing."[23]

Gleason did not publish much as a junior fellow and was certainly freer than most of his mathematical peers to follow his imagination, wherever it happened to lead him, but he still made enough of a mark during his four years in the Society of Fellows to convince Harvard to keep him around. Gleason joined the faculty as an assistant professor in 1950—despite the fact that he had not amassed a single graduate credit, much less earned a Ph.D.—and he eventually became the Hollis Professor, the oldest endowed scientific professorship in the country, dating back to 1727. Gleason attributes his appointment to a lecture he gave to the Harvard Math Club on the theory of the mathematical game Nim, which he got interested in during his term in the Navy after the war had ended and the urgency of their mission had eased up. He then worked out a complete theory of the game, not knowing that two others had worked out a similar theory a decade before, though he still felt—upon reading the earlier work—that his approach was "more interesting." As for the talk he delivered on the subject at Harvard, some four years later, he felt "it went over very very well and attracted a lot of attention."[24] While that may be true, one can be sure that Gleason was not hired solely on the basis of his Nim musings.

One of the areas he began to work on was discrete mathematics, which in many ways followed the kind of work he was doing in cryptanalysis. In essence, discrete mathematics is about counting things. "There are only twenty-five other things you can make the letter 'A' into, and only twenty-five other things you can make the letter 'B' into, and so forth," explains Bolker. "This is how discrete mathematics comes into code breaking: you can count the number of possible ways of putting something into code. To break the code, you have to find the one way—out of all the possibilities—that's being used by your enemy, which is basically a discrete [mathematics] problem."[25]

Of course, discrete mathematics does not have to have anything to do with codes per se; it deals with mathematical structures that are discrete (like integers) rather than continuous (like real numbers). The four-color problem (discussed in earlier chapters)—which concerns the number of colors needed so that for any map drawn in a plane, no two adjacent sections have the same color—is an example of a problem in discrete mathematics. Gleason did not devote much effort to that problem, although one of his graduate students, Walter Stromquist, wrote his thesis on the subject. Gleason did, however, become fascinated with Ramsey theory, which also relates to counting things but is, more specifically, about finding order and organized substructures among seemingly disordered structures. This theory was named after the British mathematician Frank Ramsey, who developed it in the late 1920s, just before his untimely death (from liver disease) at the age of twenty-six.

Problems in Ramsey theory can be posed in various ways. For instance, you can ask what is the minimum number of people you have to invite to a party to ensure that either three of them will know each other or three of them will be complete strangers. That number, which turns out to be six in this case, is the so-called Ramsey number—the minimum needed to guarantee a specific kind of structure within the larger group. These problems can quickly become very complicated and computationally challenging. In a paper regarded as a classic in the development of Ramsey theory, Gleason and his former OP-20-G colleague Robert E. Greenwood, a mathematician who taught at the University of Texas at Austin, established the value of several new Ramsey numbers, while placing upper and lower bounds on others.[26] They proved that $R(4,4) = 18$, meaning (in party terms) that you need to invite eighteen people to ensure that at least four of them are either strangers or know each other. They also proved that $R(3,3,3) = 17$, meaning that you need to invite seventeen people to a party to ensure that three will be friends, three will be enemies, or three will have neutral feelings toward the others.[27]

"Gleason told me he had spent a great deal of time looking for other exact values," recalled his former graduate student Joel Spencer, now at the Courant Institute of Mathematical Sciences, who pursued a Ph.D. thesis related to Ramsey numbers. "Since I knew of his legendary calculatory powers," Spencer said, he decided to devote his energies in other directions. That turned out to be a wise choice, he wrote in 2009, since "despite great efforts and high speed computers, only a few other values are known today."[28] In fact, it took about forty years for the next nontrivial Ramsey number to be established, $R(4,5) = 25$.[29] Determining the value of $R(5,5)$ is well beyond our current computational abilities. And if aliens landed on Earth and asked us to tell them the value of $R(6,6)$ or they would destroy the planet, our only recourse, according to the renowned mathematician Paul Erdös, would be "to destroy the aliens."[30]

Working out Ramsey numbers, and firming up the underlying theory, was a stimulating exercise for Gleason—and one that utilized his tremendous powers of calculation—but it is not what most people remember him for. During his years as a junior fellow, he was drawn to a problem of greater gravitas, Hilbert's fifth problem, which George Mackey discussed in a course that Gleason took in 1947 or 1948. "Andy already knew about the problem but—for reasons he says he doesn't understand—only began serious work on it after this point," said Mackey. "I well recall his telling me that he thought he could solve it and the confident air with which he said so."[31]

Gleason continued to work on this problem throughout the fellowship and after its conclusion. "I don't think anybody at Harvard was really aware of how much I was doing with the Fifth Problem," Gleason said.[32] His multiyear effort—a long-term one by Gleason's lightning-quick standards—paid off, as he ended up solving a big chunk of this famous problem more than fifty years after it was unveiled, and it is for this work that he is best known.

The fifth problem was one of twenty-three problems that David Hilbert presented in 1900 at the International Congress of Mathematicians in Paris in the hopes of advancing mathematical research. At the time, Hilbert was one of the most influential mathematicians in the world, and he was also the mathematics chair at the University of Göttingen, which was widely regarded as the most influential department in the world. The problems he chose "should be difficult in order to entice us, yet not completely inaccessible, lest it mock at our efforts," he said. And the ultimate value of these problems, which would be impossible to ascertain in advance, "depends upon the gain which science obtains from the problem."[33]

Since the question Hilbert initially posed could have been answered by a simple "no," mathematicians eventually rephrased the fifth problem to a form that could not be dispensed with so readily: "Is every locally Euclidean group a Lie group?"[34] Even though the entire problem can be summed up succinctly, in language that does not sound particularly intimidating, it is still rather difficult for nonmathematicians to grasp. The following is an admittedly simplified discussion of the problem and its eventual resolution.

The fifth problem is perhaps most readily understood in terms of topology, which was the context in which it was initially proposed, even though mathematicians later came to view it in much more general and abstract terms. Hilbert was originally thinking about groups of symmetry transformations of a manifold. We take up the latter term first before getting to transformations. A manifold is a topological space that is locally Euclidean, meaning that every point lies in a neighborhood that resembles flat space. For example, the surface of Earth is a sphere, which is also a manifold. When viewed from space, we can see that it is curved, but a tiny patch of the surface looks flat, and every tiny patch connects smoothly with the patches adjacent to it—all stitched together to make something akin to a patchwork quilt. If you look at the entire globe, the longitudinal lines intersect at the North and South Poles. But if you zoom into a very small section of this surface, such as the city of Manhattan, everything looks Euclidean again, and as a consequence, parallel lines and parallel streets do not intersect (which is desirable from a traffic management standpoint).

Rotation is a kind of symmetry transformation that a sphere enjoys. What that means, in simplest terms, is that you can rotate a sphere in any direction—and to any degree—around its center, and it still looks the same.

The same arguments apply to a circle, a one-dimensional manifold that is even simpler than a sphere. It is also a simple example of a Lie group (named after the nineteenth-century Norwegian mathematician Sophus Lie). A circle is not a Euclidean space, but it is locally Euclidean in the sense that every tiny part of the circle looks like a piece of a line. But in what sense is a circle a group? Well, for one thing, it consists of a set of points that are a specific distance, r, from the origin, and for the sake of simplicity, we will assume that in this case r equals 1. You can associate every point on the circle with an angle from 0 to 2π (or, if you prefer, from 0 to 360 degrees). This group has an operation, namely, addition: you can add two angles together, such as $\frac{1}{4}\pi$ and $\frac{2}{4}\pi$, and get a new angle, $\frac{3}{4}\pi$, that corresponds to a different point on the circle. If two angles add up to more than π—say, $\frac{3}{4}\pi + \frac{3}{4}\pi = 1\frac{1}{2}\pi$—that brings you to the same point on the circle as $\frac{1}{2}\pi$, which is an example of "modular arithmetic" (just as a standard clock cycle starts anew every twelve hours so that normally we do not get to 16 o'clock, 27 o'clock, or 39 o'clock). This operation is associative: $(\frac{1}{4}\pi + \frac{1}{2}\pi) + \frac{1}{4}\pi = \frac{1}{4}\pi + (\frac{1}{2}\pi + \frac{1}{4}\pi)$. There is an identity element—doing nothing, which is equivalent to rotating by 0π—and there is an inverse element: if you rotate by $\frac{1}{4}\pi$, you can reverse that by rotating in the opposite direction (by $-\frac{1}{4}\pi$), which takes you back to where you started. The group can be thought of as the circle (or manifold) itself or the set of all possible rotations of an arm of radius 1 affixed to the origin. And these rotations, and the various possible combinations of rotations, can be described by a function—albeit a simple function in this case.

A Lie group, by definition, is a topological group that is locally Euclidean—like the sphere and circle just discussed—which is another way of saying that it is a manifold. But a Lie group has another property that is seemingly more restrictive: the manifold must be "smooth," meaning that it has no sharp peaks or corners so that nice smooth tangents can be drawn at every point, which is the case with a sphere and circle. Saying that a manifold is smooth is equivalent to saying that it is infinitely differentiable. For each tiny patch of the manifold, there is a map or function that takes every point on this patch to a section of Euclidean space of the same dimension. A patch of the sphere would thus map—in a continuous, one-to-one fashion that leaves no gaps or tears—to and from a corresponding patch of the plane (also known as two-dimensional Euclidean space). Because the manifold itself is smooth, the function that goes from it to Euclidean space—or from Euclidean space back to the manifold—must be smooth as well, meaning that you can take its derivative an unlimited number of times.

Returning to Hilbert's proposition, a Lie group, as stated above, is locally Euclidean. What Hilbert wanted to know was whether the converse is true: Is every locally Euclidean group a Lie group? In other words, if you insist on your group being locally Euclidean, do you get something extra (this added feature of differentiability) thrown in for free? As Gleason put it, getting an affirmative answer to that question—a positive solution to Hilbert's problem—"shows that you can't have a little bit of orderliness without getting a lot of orderliness."[35]

That, at least, was what he set out to prove, but he was not starting from scratch. Hilbert had issued the challenge some fifty years earlier, and in the meantime, other mathematicians—including L. E. J. Brouwer, Béla Kerékjártó, John von Neumann, Lev Pontryagin, and Claude Chevalley—had made important contributions, solving special cases of the fifth problem, such as the cases for one, two, three, and four dimensions. Gleason was after a much more general solution, valid in any finite dimension. The trick was figuring out a way to proceed.

"It turns out to be difficult to draw useful conclusions about a topological group from the assumption that it is locally Euclidean," wrote Gleason's first Ph.D. student, Richard Palais, who is now at the University of California, Irvine. "So the strategy that Gleason devised for settling the Fifth Problem was to look for a . . . 'bridge condition' and use it in a two-pronged attack: on the one hand show that a topological group that satisfies this condition is a Lie group, and on the other show that a locally Euclidean group satisfies the condition. If these two propositions can be proved, then the positive solution of the Fifth Problem follows—and even a little bit more."[36]

The bridge used here was the condition of no small subgroups (NSS), which was a term coined by Irving Kaplansky, who left Harvard to joined Mac Lane's Applied Mathematics Group during World War II, relocating to the University of Chicago after the war. A subgroup, as the name implies, is a subset of a group that also satisfies the definition of a group: every subgroup has an identity element and an inverse element; the product of two elements in the subgroup is always an element in the subgroup; and the operation of taking a product must obey the associative law.

To get a crude notion of what NSS is all about, it is probably easiest to start by thinking about topological *spaces* rather than topological groups. If P is a point in the plane, then points close to P constitute a "neighborhood." We could define our neighborhood more precisely by saying, for instance, that it includes all the points less than or equal to the distance r from P, where r is a positive number. The neighborhoods we are talking about here are all disks, and we can construct different neighborhoods by varying the value of r. But in this case, the intersection of all the possible neighborhoods of P that we can draw is just P, which means that nothing is arbitrarily close to P except P itself.

The NSS condition is analogous to this, only it pertains to topological groups (rather than topological spaces) and their subgroups. So instead of talking about a point P in the plane, we talk about a group consisting of transformations that act on the plane. The identity element is the transformation that leaves P alone, whereas other transformations in the "neighborhood" of the identity might move P just a little bit. And just as we talked about neighborhoods around P, we can also talk about neighborhoods around the identity. NSS, Palais explains, "refers to a topological group without *arbitrarily* small subgroups—i.e., one having a neighborhood around the identity that includes no subgroup except the trivial group."[37] And the trivial group is a group consisting of just a single element—the identity itself. If, on the other hand, the neighborhood contains a subgroup with more than one element (i.e., not just the identity), NSS does not apply. In fact, adds Palais, "that's precisely the kind of situation that NSS screens out."[38]

Using NSS as a "bridge," Gleason achieved the first part of his "two-pronged" strategy, proving that a topological group that satisfies NSS must be a Lie group. He intended to tackle the second part next, showing that a locally Euclidean group satisfies NSS, thereby sealing the proof. But someone else got there first.[39]

In February 1952, Gleason discussed his result with Deane Montgomery, a professor at the Institute for Advanced Study in Princeton.

Montgomery, who had solved the three-dimensional case of the fifth problem in 1948, was then finishing up his proof of the four-dimensional case with his long-term collaborator, Leo Zippin of Queens College. (Their names were so closely linked, in fact, that Zippin was once given a name tag at a conference that read: "Montgomery Zippin.")[40] As soon as Montgomery heard what Gleason had achieved, he felt confident that he could solve the remaining part of the fifth problem.[41] He and Zippin made use of Gleason's result to fill in the missing piece of the Hilbert puzzle. Working virtually nonstop, they submitted their proof to the *Annals of Mathematics* on March 28, 1952.[42] It was published in the September 1952 issue of the *Annals,* immediately following Gleason's paper.[43] "Together these results constitute an affirmative solution to Hilbert's Fifth Problem," Gleason wrote.[44]

In Palais's opinion, "the much more difficult part of the problem was the first part, which Gleason proved."[45] Montgomery apparently concurred, claiming that "the most ingenious part" of the combined proof lay within Gleason's paper.[46]

As far as Palais ever heard, Gleason never expressed any dismay over the fact that Montgomery and Zippin got there first.[47] If anything, Gleason said pretty much the opposite, acknowledging that his approach "didn't even take account of the local Euclideanness [of these groups]. That's why it didn't solve the problem by itself." He further noted that "it is ridiculous to assume that any one person can solve a serious problem. That almost never happens."[48]

But to a large extent, that did happen in this case, as Gleason solved his part of the puzzle working entirely on his own. And after more than fifty years of effort by many distinguished mathematicians, one could definitively answer the challenge posed by Hilbert and subsequently reformulated by his peers: "Yes, every locally Euclidean group is indeed a Lie group." But what more could one say? In presenting his twenty-three questions, Hilbert had hoped to open new doors in mathematics leading to untold treasures. In this case, sadly, that did not really happen, despite the intricate arguments advanced by Gleason, Montgomery, Zippin, and their predecessors (as well as others who came later, such as Hidehiko Yamabe, who generalized Gleason's result).

Speaking of Hilbert's fifth some three decades after it was solved, the French mathematician Jean-Pierre Serre said: "When I was a young topologist, that was a problem that I really wanted to solve, but I could get nowhere. It was Gleason and Montgomery-Zippin who solved it, and their solution all but killed the problem. What else is there to find in this direction?"[49]

Palais agrees. "It's nothing against Gleason. What he did was still a tremendous tour de force. But the main effect was that it stopped other people from working on it."[50] Fortunately, it did not stop Gleason from working on other problems, because solving problems was one thing he loved to do. Nor, it seems, could he keep himself from taking on problems—even when it was not apparent to other people that there was a problem to be solved. At an annual math department picnic, for example, Gleason amused himself by figuring out how to do cube roots on an abacus.[51] When Gleason and a colleague were bumped to first-class seats on a flight, rather than reveling in the unexpected food and drink that he could partake of, Gleason instead listened to the pilots' radio communiqués in order to compute the amount of fuel being loaded onto the plane.[52] Even an outing to Fenway Park, where Gleason and friends took in a Red Sox game, posed unexpected math challenges. Prior to the game, Gleason reported "the results of an analysis he had just conducted that morning proving that the standings in each of the four divisions could easily be explained totally on the basis of chance phenomena rather than ability or the lack of it for any of the teams," recalled Sheldon Gordon, who collaborated with Gleason on several calculus textbooks.[53]

Having observed this sort of behavior for many decades, Bolker recognized what was obvious to most people who knew Gleason: He was "a problem solver more than a theory builder. He liked hard problems, like Hilbert's Fifth . . . Others less deep interested him no less."[54] He also was curious about history and often wondered what the ancient Greek mathematicians knew and did not know if one were to translate their ideas into contemporary terms. Owing to his longstanding interest in this area, Gleason took on a classic problem in geometry regarding the construction of regular polygons—a polygon whose sides and angles are all equal. The Greeks knew, for example, that one could construct an equilateral triangle, square, pentagon, and many other regular polygons, as well as polygons derived from these, strictly by using a straightedge and compass, but they were not able to use this method to construct, say, a seven- or thirteen-sided regular polygon. Carl Friedrich Gauss proved more than two hundred years ago that it is impossible to make a seven-sided regular polygon in that way. Gleason proved that you could construct seven- and thirteen-sided regular polygons with a straightedge and compass, provided that you also have the ability to trisect an angle. Gleason's major accomplishment, however, was in identifying *all* the regular polygons that could be constructed with a straightedge, a compass, and the capability for angle trisection.[55] He also established definitively (though it had been known before) that having the means to trisect angles does not enable you to solve another classic problem, known as "doubling the cube," that had stymied the Greeks: use a straightedge and compass to determine the length of a side of a cube that would have twice the volume of a given cube. We cannot do it any better than the Greeks could more than two thousand years ago, but now we at least know that it is impossible.

"This was a very pretty piece of work on Gleason's part that was, in some sense, insignificant," Bolker said. "You could know upfront that there was no possible way that this could advance the field of mathematics, but he didn't care. It was still a pretty problem."[56] And it also gave Gleason a chance to use the word "triskaidecagon"—a thirteen-sided regular polygon—which does not come up much in common parlance. This, Bolker surmised, might have been part of his motivation.[57]

Of course, Gleason did not shy away from problems of greater consequence. He worked, for example, on the Riemann hypothesis, which has defied the world's best mathematicians for more than 150 years (himself included on a long list of those who have tried before and since). He also proved an important theorem related to quantum mechanics, now known as "Gleason's theorem," which he got going on—as with Hilbert's fifth problem—thanks to some vigorous encouragement from Mackey.

Before talking about Gleason's theorem and the Mackey conjecture that led to it, let us say a little bit about Mackey's contributions to the mathematical foundations of quantum mechanics—an area of longstanding interest to him. Mackey, in turn, built on the previous work of his Ph.D. supervisor, Marshall Stone, and of John von Neumann. Stone and von Neumann worked independently to create a mathematical framework for the Heisenberg uncertainty principle, which states that the accuracy of a particle's momentum measurement is inversely proportional to the accuracy of the particle's position measurement. The Stone–von Neumann theorem reframes this relation in terms of "commutativity": in quantum mechanics, every "observable," such as momentum and position, is replaced by an operator, and these two operators do not commute. The order in which you measure momentum and position does matter since measuring momentum will affect your subsequent position measurement and vice versa.

In what is now called the Stone–von Neumann–Mackey theorem, Mackey provided a more abstract version of the above relations, which worked in a much more general situation that had nothing to do with momentum or position. "Stone and von Neumann approached the problem solely in the quantum mechanical context," says Yale mathematician Roger Evans Howe (whose adviser Calvin Moore was Mackey's student). "They were doing mathematical physics. Mackey took it out of the physics context and put it in a general mathematical context. Andre Weil subsequently observed that some special cases of Mackey's theorem were relevant to understanding some of the deepest results of number theory of the first half of the 20th century."[58]

"Unlike many mathematicians, he [Mackey] was not trying to solve all the problems of the physicists, nor was he trying to tell the physicists what they were or should be doing," added UCLA mathematician Veeravalli S. Varadarajan. "He was very much more modest, content to understand and interpret the physicists' conception of the world from a mathematician's point of view."[59]

In closely related work that Gleason eventually got involved in, Mackey thought deeply about Born's rule, a key tenet of quantum mechanics named after, and derived by, the physicist Max Born. Born's rule (also known as Born's law) provides the probability that a measurement of, say, a particle's position will yield a particular result. More specifically, Born proposed that the probability of finding an object at a certain time and place is equal to the square of its wave function, *psi*. Built into Born's method was the assumption that in quantum mechanics states are described by unit vectors, and probabilities are computed from these vectors. (They are called "unit vectors"—of length one—because probabilities can never exceed 1.)

Mackey wanted to show from first principles that representing states by unit vectors—and using those vectors, in turn, to calculate probabilities—was mathematically justifiable. At stake was a broader issue known as the hidden variable problem: "In quantum mechanics, you only calculate probabilities," explains Varadarajan. "Is that because our knowledge of a state is incomplete, due to 'hidden variables' of which we know nothing, or is it because nature obeys quantum mechanics? Von Neumann showed that it is not the incompleteness of our knowledge that leads to the statistical basis of quantum mechanics but rather something intrinsic to nature itself. However, von Neumann's axioms were very restrictive. Mackey wanted to reach the same goal, showing that states could be described by unit vectors, without the help of those [overly] strong axioms."[60]

In essence, Mackey was hoping to get rid of any unnecessary conditions and reformulate the idea in the most general, and minimal, way possible. He recast the problem in the precise mathematical form of a conjecture. "A positive answer to Mackey's question would show that the Born rule follows from his rather simple axioms, and thus, given these weak postulates, Born's rule is not ad hoc but inevitable," wrote Berkeley mathematician Paul R. Chernoff, a former Gleason Ph.D. student.[61] A proof of the conjecture, in other words, would argue against the existence of hidden variables in quantum mechanics, thus showing that quantum theory offers a fundamental departure from the classical way of viewing the world.

The main problem, from Mackey's point of view, is that he did not know what to do with his beautiful mathematical formulation. "I saw no way of proving the theorem" and, in 1956, "mentioned my conjecture to Andy [Gleason] in the spirit of telling him what was on my mind," Mackey said. "I did not think he would be interested in working on it, and I do not believe that I suggested that he do so. However, it caught his fancy and judging from his progress reports, he worked on it with great intensity for some time until he finally solved it. In my opinion, this work is one of his greatest achievements"—one that has had important applications in physics, even more so than in math— "and I am proud of my part in it, however inadvertent it may have been."[62]

"To Mackey's surprise, Andy was seized by the problem with intense ferocity," Chernoff added.[63] Gleason broke it down into several parts, first proving the conjecture in three dimensions and then extending that result into all higher dimensions, including the infinite-dimensional generalization of Euclidean space known as Hilbert space (although Gleason's theorem applies only to Hilbert spaces of the "separable" variety).[64] Mackey and Gleason were good friends, and Gleason regarded Mackey as his mentor and Ph.D. supervisor—despite the fact that he never got a Ph.D.—but they never collaborated on any papers.

This, in fact, was the only time that Gleason proved a theorem posed by Mackey, so it might be fair to say that their professional careers intersected, quite literally, in Hilbert space.

Varadarajan calls Gleason's proof of Mackey's conjecture "a team victory. Without Gleason, Mackey never would have proved it. Without Mackey, on the other hand, Gleason wouldn't have known what to prove."[65]

"Gleason and Mackey were very different," says Howe. "Mackey was very slow, but he had an internal vision that he pursued with tremendous determination over the years."[66] Gleason, on the other hand, worked with dazzling speed, possessing "the metabolism of a hummingbird," as one prominent physicist put it.[67] While Mackey was a builder of theories, adds Howe, "Gleason was a brilliant guy who could figure things out quickly. He would respond to external challenges, but there didn't seem to be something egging him on inside. In mathematics, we need both kinds of people."[68]

The two friends met frequently for long conversations about mathematics, and Mackey admitted to being awed by his younger colleague's swiftness both in grasping new concepts and in solving problems. "Indeed, the one flaw in our discussions was Andy's tendency to go too fast for me to follow," said Mackey. "I must confess that I often found myself nodding vaguely as though I understood rather than interfere with some of his lengthy rapid utterances."[69]

Mackey, despite those words, was no slouch either, as he, too (like Gleason), was a top five Putnam Competition winner during his undergraduate years at Rice University—a feat that earned him a full scholarship to Harvard for graduate school. Mackey gained "peace of mind," however, once he stopped "regarding Andy as a dangerous younger rival whom I had to outdo."[70] Rather than trying to win a race, Mackey concentrated on his own strength, which was his ability to think deeply on a subject—for years or decades at a stretch—working day and night, with monklike devotion. He maintained a steady focus on the big picture, concerning himself more with the overall structure of a given subject than with the details of individual proofs. "I want to use a telescope, not a microscope," he said.[71]

"He was a scholar in the truest sense, his entire life dedicated to mathematics," said University of Warwick mathematician Caroline Series, a former Mackey Ph.D. student. "He lived a life of extraordinary self-discipline and regularity, timing his walk to his office like clockwork and managing—how one cannot imagine—to avoid teaching in the mornings, this prime time being devoted to research."[72]

To minimize distractions, Mackey had no phone in either his Harvard office or his home study. "The visual arts left him unmoved; instead he saw great beauty in mathematics," his daughter, Ann Mackey, wrote. "New insights thrilled him, and he would emerge from his study giddy with excitement about a new idea." However, mostly he did not emerge from his study, spending most of his days holed up there, working as much as he could, coming out only for meals. "If I came upstairs to look for him, I could count on seeing him slumped in his chair, with clipboard in hand, lost in thought," his daughter added.[73] Mackey maintained a written record of every minute he worked—spending forty-five minutes of an hour on his research and fifteen reading in another field to "cleanse his intellectual palate"—all part of his effort to wring the maximum possible out of each and every day.[74]

In the end, Mackey produced an impressive body of work and left behind schools of followers to help carry out his vision. University of Colorado at Boulder mathematician Judith Packer, a former Ph.D. student of his at Harvard, considers Mackey "one of the great mathematicians and mathematical characters of the past century."[75] He is probably best known for his work in two key, though overlapping, areas: representation theory, in which he is considered a giant, and the mathematical foundations of physics, including his aforementioned contributions to quantum mechanics, for which he developed and applied ideas from representation theory. "His ability to strip things down to their essential mathematical structure put a hugely influential stamp on generations of mathematicians and physicists," noted Series.[76]

In representation theory—the field for which Mackey is probably best known—a "representation" of a group is a way of representing each and every element of the group by a matrix. Matrices, in turn, can be used to represent linear transformations of vector spaces—the latter being a collection of vectors that can be added together or multiplied by numbers (or "scalars"). Putting those notions together, Roger Howe has described a representation as "groups acting on vector spaces." The basic idea, he adds, "is to take any abstract group and find out all the ways it can act by linear transformation on a vector space."[77]

One reason that representation theory has proved so useful is that an abstract group, which might have been hard to work with in its original form, can be recast, instead, as a group of matrices—concrete objects that are relatively easy to manipulate. Group operations, such as multiplication and addition, are thus given by matrix multiplication and addition, and problems in abstract algebra can be converted into problems in linear algebra—an area that is reasonably well understood.

Mackey restricted himself, for the most part, to the study of unitary representations—representations of groups whose elements consist of unitary transformations, which is to say, transformations that preserve distance. Perhaps the simplest example of a unitary (and hence distance-preserving) transformation is the rotation of the plane, although the term "unitary" implies complex vector spaces in a complex, rather than Euclidean, plane.

Returning for the moment to simpler, and more intuitive, Euclidean space, one can consider a cube sitting on a table and all the ways it can be picked up and put back in an identical-looking position. First, there are six faces it can be lying on, and any vertex touching the table can assume one of four positions. So that is 24 (6 × 4) possibilities, and if you allow reflections, there are 48 (6 × 4 × 2) possible ways of picking up the cube and putting it down so that it ends up looking the same. This group of forty-eight symmetries—each element acting on three-dimensional space—is a three-dimensional representation of a group of forty-eight elements.

Inspired by the physicist Eugene Wigner, who helped bridge the gap between group representations and physics, Mackey realized that representation theory was almost perfectly suited to quantum mechanics. One reason the fit is so good is that a group can be represented by elements (matrices) that act on vector spaces. This meshes with quantum mechanics, where states are described by unit vectors whose lengths remain invariant under transformations. Consequently, Mackey was able to put this methodology to good use in generalizing the Stone–von Neumann theorem and in developing the conjecture related to Born's rule that Gleason proved. "Ultimately I found that the mathematical theory I had been developing [the theory of unitary group representations] was almost the ideal tool for understanding the whole structure of quantum mechanics," Mackey wrote in a letter to his daughter's friend, Stephanie Singer. "Still later I became interested in the fact that this same mathematical theory . . . has extensive applications to the theory of numbers and began to learn number theory (of which I was absolutely ignorant) and to develop its connections with unitary group representations."[78]

One thing Mackey is famous for in representation theory is the so-called Mackey machine. Although his contribution is really quite technical, Varadarajan describes the "machine" as "a template for creating representations of groups."[79] To try to get a very rough sense of this concept, let us start by supposing a particular group is made up of various pieces—subgroups, normal subgroups (which are special kinds of subgroups), and quotient groups (which are made from a normal subgroup). "If you know the representation of the pieces—if you know the representations of the normal subgroup and the quotient subgroup," Howe explains, "the Mackey machine tells you how to construct the representation of the whole group." The Mackey machine highlights the importance of *induced representations*—a concept that Mackey brought to the fore—which involve a procedure for constructing a representation of a group from a representation of one of its subgroups. Induced representations are also central to the Mackey imprimitivity theorem, one of the main theorems that came out of the Mackey machine, which has had important applications in quantum theory, including the extension of the Stone–von Neumann theorem.[80]

Mackey also applied representation theory, with great success, to questions in ergodic theory, developing in the process "his startling and fundamental generalization of ordinary groups, which he called *virtual groups*," said his friend and fellow Harvard mathematician David Mumford, who credits Mackey with showing him "for the first time the beauty of the world of mathematics"[81] and leading him "on a yellow-brick road to more and more amazing places."[82]

Mackey devoted his entire career to pure mathematics, never deviating from his course until his health failed him in his final years—the clipboard, which "traveled with him around the world," never far from his side.[83] He died of pneumonia in 2006, at the age of ninety, at his home in Belmont, Massachusetts.

Gleason outlived his mentor by two years and never stopped thinking about mathematics until his death from surgical complications in 2008. Like Mackey, he always carried a clipboard with him, "even around the house," his wife, the psychologist Jean Berko Gleason, recalled, "and filled sheets of paper with ideas and mysterious (to me) numbers. When he was in the hospital during his last weeks, visitors found him thinking deeply about new problems."[84]

Unlike Mackey, Gleason did not focus exclusively on pure mathematics in the latter stages of his career. He served as president of the

American Mathematical Society in 1981–82 and, over time, became increasingly interested in mathematics education. He wrote his first and only solo textbook, *Fundamentals of Abstract Analysis,* in 1966, which was around the same time he became involved in reforming the K–12 mathematics curriculum in the United States. In his review of *Fundamentals,* the French mathematician Jean Dieudonné stated that Gleason's book, in contrast to most texts, provided a genuine sense of what mathematics is really about:

> Every working mathematician, of course, knows the difference between a lifeless chain of formalized propositions and the "feeling" one has (or tries to get) of a mathematical theory, and will probably agree that helping the student to reach that "inside" view is the ultimate goal of mathematical education, but he will usually give up any attempt at successfully doing this except through oral teaching. The originality of the author is that he has tried to attain that goal in a textbook, and in the reviewer's opinion, he has succeeded remarkably well in this all but impossible task.[85]

A series of calculus textbooks that Gleason coauthored—intended for high school students and other learners—was published in the 1990s and all the way up through 2010. But his interest in mathematics education did not, by any means, come to him late in life. Speaking of his college days at Harvard in the 1980s, MIT mathematician Bjorn Poonen noted that "Andy struck me as someone genuinely interested in helping younger mathematicians develop. When I was an undergraduate, he volunteered an hour or two of his time each week for an informal meeting in his office with a group consisting of me and one or two other math undergrads to discuss whatever mathematics was on our minds."[86]

Joel Spencer, who got his Ph.D. in 1970 under Gleason's supervision, spoke of his "good fortune" at being Gleason's teaching assistant, fondly describing their conversations before class. "Andy would discuss the mathematics of the lecture he was about to give," Spencer said. "He was at ease and spoke of the importance and interrelationships of the various theorems and proofs. My contributions were minimal but I listened with rapt attention. It was in those moments that I learned what being a mathematician was all about."[87]

While thinking about geometry, Gleason once told his friend and former student Ethan Bolker "that he'd give a lot for one good look at the fourth dimension."[88] One can only hope that, upon leaving our mundane three-dimensional world in 2008, Gleason finally attained such a view—and liked what he saw.

7

THE EUROPEANS: OSCAR ZARISKI, RICHARD BRAUER, AND RAOUL BOTT

In the late 1930s and early 1940s, before and during World War II, a wave of European mathematicians, most of whom were Jewish, migrated to the United States. Although the total number of mathematicians was not especially large—an estimated 120 to 150 émigrés had arrived by the end of the war—many of these individuals were scholars of the first rank, and their influence was widely felt on the American mathematical scene.[1]

As discussed in Chapter 3, Harvard's math department was mostly untouched by the influx of European refugees until after the war. Lars Ahlfors, who was Finnish but not Jewish, visited the department from 1935 to 1938 before accepting a permanent faculty position in 1946—a year after World War II ended. Following a temporary stint at Harvard from 1940 to 1941, the Russian-born Oscar Zariski rejoined the department in 1947 and stayed there for the remainder of his long and productive career. Richard Brauer did not join the Harvard faculty until 1952 but left his native Germany in 1933 when Hitler came to power and Jewish academicians like him were forced to give up their teaching positions. The Hungarian-born Raoul Bott, who was partially of Jewish descent (but of non-Jewish upbringing), immigrated to Canada in 1938 and did not make his way to Cambridge until 1959. But all three of these Europeans—Zariski, Brauer, and Bott—ended up making a huge mark at Harvard and in their respective fields, which consisted primarily of algebraic geometry, group theory, and topology.

Oscar Zariski

Although Zariski was notable, on the one hand, simply for being the first Jew to receive tenure in the Harvard mathematics department, he had a huge impact on mathematics that had nothing to do with his religious persuasion. (In fact, he considered himself an atheist.) Zariski was largely responsible for overhauling algebraic geometry and putting it on a firmer, more algebraic footing than it had been in the past. By bolstering the tools of algebraic geometry, Zariski and his colleagues (including the French mathematician André Weil) were able to go further, and penetrate more deeply into the core of mathematics, than their predecessors could manage. Over the span of nearly a half century, he—as much as any other practitioner of his era—helped shape the development of this field, laying the groundwork for many advances in the decades to come.

While most people consider mathematics hard enough on its own, Zariski had to overcome countless obstacles and hardships in order to carve out a career in his chosen field, which is something he yearned for from an early age. It took single-minded, unflappable determination on his part to pursue mathematics amidst the chaos that often surrounded him—be it from World War I, the Russian Revolution, the rise of Nazism and Italian Fascism, World War II, the Holocaust, and the infirmities that beset him in old age. Yet, somehow, Zariski persisted in the face of all this, achieving the distinction to which we now pay tribute.

Originally named Ascher Zaritsky, he changed that to the more Italian-sounding Oscar Zariski (at the suggestion of one of his Italian mentors) after becoming a graduate student in Rome. Zariski was born in 1899 in the city of Kobryn, which was situated in a part of Russia that is now called Belarus. As a child, Zariski spent hours doing math problems with his older brother, Moses, whom Oscar quickly surpassed. While still in his teens, Zariski described his appreciation of mathematics in a diary entry: "You begin with some question . . . , and step by step you witness the wonderful functioning of your own intellect. You stumble on new problems, which in their further development lead you to new results . . . Well, to put it briefly: In mathematics I feel absolutely sure of myself."[2]

When fighting from World War I broke out in the Kobryn area, Zariski moved to Kiev, enrolling in the University of Kiev in 1918. He was unable to enter the mathematics program because those spots had already been filled, so he joined the philosophy program instead, studying mathematics—mostly algebra and number theory—on the side. Yet he was already certain, even as a teenager, of his love for mathematics. Writing in his diary that year as a student in Kiev, he said: "I feel exceptionally sure of that darling old lady, and I also feel exceptionally certain that she will not betray me, because I feel inside me the presence of a mathematical talent."[3]

The Russian Revolution broke out in 1918, and intermittent waves of Bolshevik, White Russian, and Ukrainian troops swept through the city, making it a less than ideal setting for academic pursuits. A year later Zariski was wounded, taking a piece of shrapnel in his leg when he was accidentally caught in the crossfire between Bolsheviks and Ukrainians. By 1921, Zariski decided it was time to leave strife-ridden Kiev and its university, which by this stage was barely functioning.

He chose to go to Italy, despite the fact that he had no friends or relatives there and barely any funds for transport or living expenses. Looking back years later, he realized how momentous that decision had been for him. "Boarding that train meant the end of a whole way of life, but . . . I knew that I was destined to do mathematics," he said. "I was so fascinated that when I looked at the sky I saw the birds whirling like numbers."[4]

Italy, and specifically Rome, turned out to be a good choice for Zariski. The cost of living was low, and foreign students got free tuition. Of perhaps greater significance was the fact that the Sapienza University of Rome, where Zariski enrolled, was the world center of algebraic geometry at that time. "I had the great fortune of finding there on the faculty three great mathematicians, whose very names now symbolize and are identified with classical algebraic geometry," Guido Castelnuovo, Federigo Enriques, and Francesco Severi. "It was inevitable that I should be attracted to that field," wrote Zariski.[5]

Had he not ended up in Rome, he may have continued his study of pure algebra. "But it's so much nicer when you have geometry combined with algebra," Zariski said.[6] As for what algebraic geometry really is, he explained that mathematicians disagree over whether it is the study of geometry by algebraic means or whether it is more about putting algebra into a geometric form. Nevertheless, most would agree that the field combines algebra and geometry in various ways, many of which involve the application of algebraic techniques to problems in geometry.

Algebraic geometry is often described as the study of geometric objects (of any dimension) that are defined by algebraic equations—specifically polynomial equations. As a simple example, take the polynomial equation $x^2 + y^2 - c = 0$, where c is a positive constant. Plotting the solution to this equation on an x-y plane traces out a familiar curve, a circle. Other polynomial equations in two variables of the form $f(x,y) = 0$—which are made by adding, subtracting, or multiplying x and y by themselves, by each other, and by constants—yield different curves. Polynomial equations of three variables, such as x, y, and z, yield so-called algebraic surfaces. The scope of algebraic geometry has expanded over the years to encompass higher-dimensional objects, too. Modern algebraic geometry allows the variables and solutions to polynomial equations to be drawn not just from real numbers but also from complex numbers, and even from more complicated "ground fields," as discussed below.

The fact that algebraic geometry is confined to polynomial equations is not a serious limitation, given that the influence of these equations is really quite pervasive. The phenomena we see in nature can, to a surprising extent, be described by polynomials. Moreover, almost every equation or function can be approximated by polynomials, which makes these equations important examples to study.

Zariski was quickly brought up to speed on broad aspects of algebraic geometry during his years in Rome. In fact, early in Zariski's first term at the university, Castelnuovo persuaded him to sign up for an algebraic geometry course intended for third-year graduate students. When Zariski mentioned that he had not mastered all the prerequisites, Castelnuovo told him to go to the library: "There are books and you can read them."[7]

The year 1924 was eventful for Zariski. He married Yole Cagli, who would remain his lifelong companion. He also got his Ph.D., three years after arriving in Rome. His dissertation was on an algebraic topic related to Galois theory, which Castelnuovo suggested. A branch of abstract algebra, Galois theory is named after the French mathematician Évariste Galois, who died in 1832 from wounds sustained in a duel. Galois was just twenty years old at the time of his death yet had already had a huge impact on mathematics. The theory for which he is remembered paved the way for using group theory to find the "roots" (or zeroes) of polynomial equations. In particular, Galois theory associates a so-called Galois group with a polynomial equation, and the properties of this group—which measures the symmetry of the equation—can reveal whether or not the equation can be solved by "radicals," meaning that the roots of the polynomial can be expressed (or solved for) in terms of sums, products, square roots, cube roots, and nth roots of the equation's coefficients.

The quadratic formula for polynomials of second degree—an example that many people encounter in high school math—provides a solution of this type. It holds that an equation of the general form $ax^2 + bx + c = 0$ has two solutions:

$$x = \frac{-b \pm \sqrt{b^2 - 4ac}}{2a}$$

Galois theory explains why there is no comparable formula for equations of degree 5 and higher.

Zariski tackled a different though somewhat related problem for his thesis, which Castelnuovo described as follows: "Take a polynomial equation in two variables, x and t, but which is linear in one variable, say in t. So the equation is of the form, $f(x) + tg(x) = 0$, where f and g are polynomials in x. If you consider that as an equation in x, then x is an algebraic function of t." Zariski's assignment was to find all equations of this kind that could be solved for x in radicals, and he showed that there are five classes of such equations, all of which are related to trigonometric or elliptic functions.[8]

In work that built on his thesis, Zariski looked at "general algebraic curves" over complex numbers, which are curves of one complex dimension or surfaces of two real dimensions. (The number of dimensions doubles in this way because a complex number, z, assumes the general form $x + yi$, where x and y are real numbers. A single complex variable thus corresponds to two real variables, so that a pair of coordinates is required to locate a solution on the real, two-dimensional plane.) Zariski proved a conjecture posed by Enriques in 1897 that a general curve of genus greater than 6 cannot be represented by an equation of the form $f(x,y) = 0$, where x and y are expressed in terms of radicals of the parameter t. "Even this early in his development, we can see his tendency to combine algebraic insights and topological ideas with the synthetic ideas of classical geometry," commented Zariski's biographer Carol Parikh.[9]

Zariski spent three years doing postdoctoral work in Rome, where he continued, among other research, to pursue questions in algebraic geometry related to Galois theory. Castelnuovo and Severi encouraged him to explore Solomon Lefschetz's innovative work in topology, which they believed represented the road forward for algebraic geometry. Zariski delved deep into Lefschetz's oeuvre, which he found stimulating, just as his mentors had suggested.

But the connection also proved instrumental on another front. Anti-Semitic policies were becoming more entrenched in Italy ever since Benito Mussolini, head of the National Fascist Party, became prime minister in 1922. Zariski decided it would be wise for him to leave Italy, and he turned to Lefschetz, who had recently joined the Princeton University faculty, for help. As a Russian-born Jew, Lefschetz had much in common with Zariski, and he helped the younger mathematician secure a research fellowship at Johns Hopkins University for the 1927–28 academic year. Before the end of that fellowship year, Johns Hopkins offered Zariski an associate professor position, which he accepted. He stayed at Johns Hopkins for almost twenty years, becoming a full professor in 1937.

"For almost ten years after I left Rome in 1927," Zariski wrote,

> I felt quite happy with the kind of "synthetic" (an adjective dear to my Italian teachers) geometric proofs that constituted the very life stream of classical algebraic geometry (Italian style). However, even during my Roman period, my algebraic tendencies were showing and were clearly perceived by Castelnuovo, who once told me: "You are here with us but are not one of us." This was said not in reproach but good-naturedly for Castelnuovo himself told me time and time again that the methods of the Italian geometric school had done all they could do, had reached a dead end, and were inadequate for further progress in the field of algebraic geometry.[10]

One of Zariski's main achievements in this decade was publishing the monograph *Algebraic Surfaces* in 1935. The treatise, which Zariski hoped would provide the definitive account of algebraic geometry to date, afforded him international recognition as an expert on the subject. The work was heavily influenced by Lefschetz's topological ideas, which had prompted Zariski to approach problems in algebraic geometry from a new perspective. As for why he focused on algebraic surfaces, Zariski explained, "In my student days in Rome, algebraic geometry was almost synonymous with the theory of algebraic surfaces. This was the topic on which my Italian teachers lectured most frequently and in which arguments and controversy were also most frequent. I felt that this would be the real testing ground for the algebraic methods which I had developed earlier."[11]

Algebraic Surfaces, along with the work leading up to it, represented a genuine turning point in Zariski's career. In the monograph he tried to encapsulate the ideas underlying the methods and proofs of the Italian geometers pertaining to the theory of surfaces. "In all probability I succeeded, but at a price," he said. "The price was my own personal loss of the geometric paradise in which I so happily had been living. I began to feel distinctly unhappy about the rigor of the original proofs I was trying to sketch (without losing in the least my admiration for the imaginative geometric spirit that permeated those proofs); I became convinced that the whole structure must be done over again by purely algebraic methods."[12]

He had gotten glimmers before that the edifice upon which the algebraic geometry of the Italian school rested "was shaky in its foundations."[13] In a 1928 paper, for example, he found a clever way of correcting an incomplete proof of Severi's that had been presented as if it were a fait accompli. Zariski had also been troubled by Enriques's cavalier attitude toward proofs in general—an attitude, he believed, that likely contributed to the eventual demise of the Roman era of algebraic geometry. "We aristocrats do not need proofs," Enriques told him. "Proofs are for you plebeians."[14] But the deeper Zariski looked, the more he found that the lack of rigor was more widespread than he had ever imagined. Many of the classical proofs were, like the aforementioned Severi effort, incomplete or imprecise at various points. This realization, which took a number of years to hit home, prompted Zariski to take it upon himself to "rebuild the foundations of the field."[15]

By then, he was convinced that the classical language of algebraic geometry was inadequate. The field would have to be rewritten in terms of modern commutative algebra, and he devoted a couple of years almost exclusively to studying the subject—much to the dismay of Johns Hopkins president Joseph Ames, who inquired as to why Zariski's publications had dropped off so and was not entirely satisfied by the answer.[16]

Zariski was fortunate at this particular juncture to have been invited to spend the 1934–35 academic year at the Institute for Advanced Study. Emmy Noether, a leading developer of algebra in that era, had recently become a lecturer at the institute after having been dismissed from the University of Göttingen for being Jewish. Zariski met regularly with Noether during that year, and he profited greatly from those exchanges.

He started taking the tools he had imported from algebra and applying them in original ways to problems in geometry. These new tools were much more abstract than the standard ones, but Zariski saw the relevance that these abstract ideas held for geometry. He would master complex concepts from algebra and somehow divine their geometric significance. From that point on, almost his entire career revolved around this undertaking.

Starting around 1937, Zariski explained, "the nature of my work underwent a radical change. It became strongly algebraic in character, both as to the methods used and as to the very formulation of the problem studied. (These problems, nevertheless, always have had, and never ceased to have in my mind, their origin and motivation in algebraic geometry.)"[17] Despite the importance that algebra held in his work, Zariski always insisted that the prime source for his ideas came from geometric intuition. "I wouldn't underestimate the influence of algebra, but I wouldn't exaggerate the influence of Emmy Noether," he said on another occasion. "I was always interested in the algebra which throws light on geometry, and I never did develop the sense for pure algebra. Never. I'm not mentally made for purely formal algebra, formal mathematics. I have too much contact with real life, and that's geometry. Geometry is the real life."[18]

In 1935, Zariski was asked to explain what this new geometry was all about. He and André Weil—the other leading figure trying to transform algebraic geometry in that era—were invited to give a series of lectures on algebraic geometry at the University of Moscow. The lectures, Zariski reported, went over well with the younger mathematicians in attendance but led to a "real revolt" among the old-school geometers. "Is this algebraic geometry?" they complained. "We've never seen such geometry."[19]

The brand of work that Zariski favored was, of course, far more abstract than what geometers were typically accustomed to, and rather than shy away from that abstraction, Zariski embraced it head-on. When a student asked him whether he should go into physics or math, Zariski told him to "choose math. It's more useless."[20]

That remark was clearly made tongue in cheek, since Zariski recognized, as well as any of his peers, how abstract notions from algebra could be indispensable to geometry. Algebra could bring greater rigor to a field that in the past had depended too heavily on intuition. "His philosophy was that when you base geometry on algebra, you can avoid being misled by geometric intuition," explains the algebraic geometer Heisuke Hironaka, a former Ph.D. student of Zariski's. "He said that when he writes algebraic geometry based on algebra, the rigor is automatic; it's unquestionably there."[21]

Relying on algebra can also help mathematicians deal with higher-dimensional objects that are not readily visualized. "Sometimes you'd like to manipulate the geometry but have no way of picturing it," Hironaka said. "But if you can represent the geometry in an algebraic form, algebra is based on equations, and we know how to manipulate equations. We don't find out how the geometry changes until we get to the final stage." This is exactly how Zariski wrote his papers, Hironaka added. "It was completely algebraic. He really changed the whole complexion of the field, making it possible to deal with higher dimensions without having to visualize them."[22]

Here is another example of the precision that algebra can bring to bear. Take the polynomial equation $(x - y)^3 + tx = 0$, which defines a smooth, differentiable curve if t is not zero. A curious thing happens, however, when t goes to zero. You are left with the equation $(x - y)^3 = 0$, the solution to which is the line $x = y$—or three lines, actually, all three sitting on top of one another. In other words, they "triple up." This situation, called a "degeneracy," is readily apparent in algebra but less obvious in geometry. If you are relying just on geometry and all you can see is a picture, the solution would appear to be a single line. Algebra, however, provides a clearer picture of what is going on, because you can see that instead of just one thing there really are three things that happen to be identical.

A new approach that both Zariski and Weil were keen on pursuing was to develop a theory of algebraic geometry that was valid for "arbitrary ground fields," operating in a space whose coordinates were not necessarily restricted to real or complex numbers. To understand what that means, we must first remind ourselves that algebraic geometry revolves around geometric objects—varieties—that are defined by polynomial equations. Solving those equations means determining which values of the variables are permissible—which combinations of, say, x, y, and z yield zeroes. But one has some latitude in deciding the "field" of numbers in which those zeroes lie. They could, for instance, be real or complex numbers, and those are exactly the kind of "continuous" systems in which geometry and topology originally developed. In the twentieth century, however, mathematicians started doing geometry in different settings, such as the field of rational numbers, which is a special case of real numbers, or integers, which are a special case of rational numbers. Mathematicians also experimented with "finite fields" containing a finite number of elements. A finite field of this sort might consist, for example, of just the numbers 0 and 1, where $0 + 1 = 1$ and $1 + 1 = 0$. Many other fields, of course, are possible.

Switching from a continuous (infinite) setting to a finite one would, no doubt, lead to an unusual kind of geometry, but one that might be useful for solving problems in number theory, for example. Zariski and Weil established a new avenue in algebraic geometry, based on finite or "arbitrary" fields, which quickly turned into a substantial growth industry.

In particular, they were interested in finding integer solutions to equations in which the polynomial does not equal zero but instead equals some multiple of a prime number p. In "mod p," as this approach is called—or in characteristic p, a term that is used almost interchangeably—the number $1 + 1 + \ldots 1$ taken p times always equals zero, where p is a prime number. In characteristic 3, for example, $1 + 2 = 0$ and $11 + 4 = 0$ because the sum is divisible by 3. Similarly, $1 + 4 = 2$ in characteristic 3, and $15 + 11 = 2$, because the sum, when divided by three, leaves a remainder of 2. Leonhard Euler and Carl Friedrich Gauss pioneered work in this area, and a century or two later, Weil and Zariski carried it much further in their extension of classical algebraic geometry, which in turn has led to important applications in number theory.

By 1940, Zariski was busy applying some of the new techniques he derived from abstract algebra to a wide range of topics in algebraic geometry. He was invited to spend the year as a visiting professor at Harvard and gladly accepted. The visit, according to Garrett Birkhoff, who taught in the department at that time, was intended to be "a kind of trial marriage, to be followed if found mutually congenial by a permanent invitation in another year or two."[23] (Garrett's father, George David Birkhoff, played a key role in getting Zariski to Harvard during his 1940–41 appointment.) Harvard had begun casting about for a geometer to replace both Julian Coolidge, who had recently retired, and William Caspar Graustein, who died in early 1941, and many felt that Zariski was the most qualified person available for the job. However, plans for making him a permanent offer were put on hold after the bombing of Pearl Harbor and America's entrance into World War II, which led to a hiring freeze at the university.

When his year at Harvard was up, Zariski returned to Johns Hopkins, where he faced a heavy course load (with eighteen hours of classes a week) that seriously cut into his research time.[24] Yet he was filled with new ideas and was now coming into a period of peak productivity, which lasted through much of his forties and into his fifties. "Many mathematicians in their forties reap the benefits of their earlier more original work, but Zariski was undoubtedly at his most daring exactly in this period," commented David Mumford, a Harvard and Brown University mathematician who earned his Ph.D. under Zariski in 1961.[25] For example, Mumford said, "Zariski's main theorem and connectedness theorem both involved using radical ideas from algebra and extracting the geometric content."[26]

Zariski's main theorem, which was published in 1943, is surely a major theorem, but it is not called "main" because it is considered the most consequential theorem he ever proved. It is called that because he proved twenty theorems in that same 1943 paper, one of which he labeled "main."[27] The theorem, which is too technical to explain in any detail here, revolves around the notion of a map. Algebraic geometers generally use maps that can be described in terms of algebra, either by polynomials or by rational functions, which are quotients of polynomials. The main theorem relates to a particular kind of map, called "quasi-finite," so named because for each point in a space Y, there are only a finite number of points in a space X that map to it. Although the subject seems arcane, the main theorem has become a basic and widely used tool in the life of an algebraic geometer. The theorem, moreover, has spawned numerous reformulations.

Zariski followed up this work with the connectedness theorem, which Mumford describes as an even "stronger version of the main theorem."[28] This proof, adds MIT mathematician Michael Artin, another former Zariski student, "was part of Zariski's long standing effort to lay the foundation for algebraic geometry and make the field more rigorous than it had been in the past."[29] Looking back, it seems pretty clear that Zariski's success in bolstering that foundation was probably more important in the long run than any individual theorem he proved.

In 1945, Zariski was invited to spend a year as an exchange professor at the University of São Paulo, where he conducted a joint seminar with Weil in which they discussed the latest approaches in algebraic geometry. At the time, Mumford notes, "people used to say that everything worth knowing about algebraic geometry was known by Zariski or Weil. Algebraic geometry was a small field then, and Zariski and Weil were leading it to a new era."[30] Both mathematicians appreciated the opportunity to exchange ideas with their leading colleague (and sometimes rival). Weil had great respect for Zariski, calling him "the only algebraic geometer whose work he trusted."[31] Zariski, for his part, appreciated the seminars, which, he said "presented me with a superlative audience consisting of one person."[32] But their relations were not always harmonious. While Zariski and Weil found each other highly stimulating, they rarely agreed. As Zariski put it, "You might just say that we were friends who fought."[33]

In 1946, Zariski accepted a research professorship at the University of Illinois at Urbana–Champaign, which offered a greatly reduced teaching load and higher salary than his Johns Hopkins job. A year later, he accepted a job offer from California Institute of Technology, only to rescind it (with Caltech's permission) when Harvard offered him a permanent position as a full professor. He came to Harvard in 1947 and never left, maintaining his association with the department for the rest of his life. "Over the next thirty years," his colleagues wrote, "he made Harvard into a world center of algebraic geometry," just as Rome had been decades earlier.[34]

Zariski brought top scholars to the department, helping with key faculty appointments, inviting superstar visiting professors such as Jean-Pierre Serre and Alexander Grothendieck, and attracting—through the strength of his work and personal aura—a talented corps of graduate students.

Grothendieck, who was based at the Institut des Hautes Études Scientifiques (IHES) near Paris, came to Harvard for the first of several visits in 1958 at the invitation of Zariski, who then chaired the department. Zariski told Grothendieck he would have to sign a pledge not to overthrow the U.S. government—a vestige of the McCarthy era that was required in order to obtain a visa in that time. Grothendieck replied that he would not sign such a pledge, adding that he would not mind going to prison as long as he had his books and students could visit him there.[35] Fortunately, Grothendieck avoided imprisonment and, instead, got to enjoy the opportunity to interact freely with Zariski and others. Overall, he found the Cambridge scene quite agreeable. "The mathematical atmosphere at Harvard is absolutely terrific, a real breath of fresh air compared with Paris, which is gloomier every year," Grothendieck wrote in a letter to his colleague Serre. "There are a good number of intelligent students here, who . . . ask for nothing more than to work on interesting problems, of which there is obviously no lack."[36]

MIT mathematician Steven Kleiman, a former Ph.D. student of Zariski's, had come to Harvard for graduate work in the early 1960s and was "dragged" to a Zariski lecture by his schoolmates during the first couple of days of classes. "It was the force of his personality that [inspired] me to study algebraic geometry, first of all, and secondly with him," Kleiman said. The field had moved on to a new phase, and Zariski was doing much to promote these new ideas—both through his own work and by bringing "the principal architects of that theory to Harvard."[37]

Joseph Lipman, another former Ph.D. student, believed that "Zariski's far-reaching influence was due in no small part to his qualities as a teacher. He projected the power which he exerted over mathematics, and which mathematics had over him. He was a vibrant embodiment of the ideal of a scholar-teacher, a model of devotion to the pursuit of the good mathematics life." His students, Lipman added, represented a fair proportion of the algebraic geometers educated in the United States during the 1950s and 1960s.[38]

After completing his undergraduate work at Columbia in 1947, Maxwell Rosenlicht headed to Harvard for graduate studies in mathematics. A Columbia professor asked Rosenlicht whether he was going to study with Zariski. When Rosenlicht asked why he should, the professor answered, "When you go to a place where a great man is doing ballet, you do ballet." Daniel E. Gorenstein, a graduate student of Zariski's, described his former adviser this way: "There was a giant working on the earth."[39]

Some years later, while lecturing in France, Zariski made a convert out of Bernard Teissier, a graduate student at IHES who subsequently came to Harvard as a research fellow. Teissier was mainly interested in number theory at the time, but the lecture prompted him to reconsider. "The mixture of geometry and algebra, expounded with Zariski's lively style, intoxicated me," he said.[40]

When Zariski described the concepts of algebraic geometry, even quite technical ones, his Harvard colleagues wrote, "you didn't have to know what these words meant to believe at once that these were not dry abstractions but came from a world which lived and breathed, which glowed with promise and with secrets."[41]

Zariski did much more, of course, than simply marshaling the troops and assembling a critical mass of algebraic geometers at Harvard. When he joined the faculty in the late 1940s, he was in complete command of his creative energies, and his research career, accordingly, was still going at full throttle. One problem that he worked hard at, as he had for the better part of a decade, related to the "resolution of singularities." This topic, one of the deepest and most fundamental in algebraic geometry, formed a central thread in his research efforts. His first proof in this area came in 1939, and he continued to pursue the problem, off and on, throughout the rest of his career.

Algebraic varieties, the mathematical objects that sit at the center of algebraic geometry, can have singularities—places, like the tip of a cone, where a surface is no longer smooth. Such a point is special (or "singular") in a geometric sense, meaning that the variety is not locally flat there. The hope is that through repeated algebraic manipulations or transformations, one can somehow smooth out or "resolve" singularities. If mathematicians can successfully deal with singularities, which essentially means making them go away, they can then apply standard techniques of analysis, or differential calculus, to the varieties—techniques that would normally break down in the face of singularities.

One way of making a singularity, explains Hironaka, is to take a geometric space, grab some part of it, and crush it to a point. Maybe the thing you crushed was originally a disk, a rectangular patch, a sphere, or something more complicated. After the compression, it may look like a point, but when you inspect it more carefully you find there is more to it—there is actually information inside. "Now, to see what is in it, you must blow it up, magnify it, and make it smooth, and then you see the whole picture," Hironaka adds. "That's the resolution of singularities."[42]

One example is a curve confined to a plane that crosses itself, such as a figure eight, with the point of intersection constituting a singularity. You can magnify that crossing all you want, but it will never go away. However, if you can add another dimension to this picture, you might be able to pull the curve apart, gently lifting it out of the plane, so that it no longer crosses itself. In so doing, you have gotten rid of the singularity.

A similar example is a cusp, described by an equation of the form $x^2 - y^3 = 0$, which has a sharp spike (or "beak") at the origin. One approach would be to take the equation for a cusp and rewrite it as the sum of two smooth (and hence nonsingular) curves that are differentiable and much easier to work with. Another approach, reminiscent of the previous example, involves pulling the two arms of the cusp vertically into three-dimensional space, so that the erstwhile singularity gets stretched out, becoming smooth. In this case, one can see that the singular curve, or cusp, is actually the projection, or shadow, of a smooth curve in higher-dimensional space.

A roller coaster offers another way of thinking about this. "A roller coaster does not have singularities—if it did, you would have a problem," explains Hironaka. "But if you look at the shadow that the roller coaster makes on the ground, you might see cusps and crossings. If you can explain a singularity as being the projection of a smooth object [onto a plane], then the computation becomes easier."[43]

Let us return to our example of a cone or, rather, two cones connected at the tip—a geometric shape that can be made by taking a line, fixed at one point, and rotating it around a circle. Although the point of intersection between the two cones looks like a hopeless singularity, that point can, in fact, be blown up and replaced with a small sphere. The set of lines that make up the cones are now detached from each other, each line hitting the sphere at a different point, and the singularity has consequently gone away. A manipulation of this sort is mathematically justifiable, even though it is not obvious (without looking in higher-dimensional space) as to why it is "legal" to do so.

The foregoing cases, nevertheless, offer just isolated examples of ways in which singularities might be resolved. What Zariski and other mathematicians yearned for, however, was something broader and more powerful—a set of general mathematical tools that could be used to resolve singularities in any variety and of any dimension. The basic idea is to take a small piece of a variety that contains a singularity and blow it up repeatedly. The hope is that if the singularity gets spread out enough, and you look at it with a powerful enough magnifying glass, then its problematic features might vanish and the object appears nonsingular instead.

"Zariski attacked this problem with a whole battery of techniques, pursuing it relentlessly over 6 papers and 200 pages," Mumford said. In the end he proved that "all algebraic varieties of dimension at most 3 ... have nonsingular models."[44] In other words, he proved that the singularities in varieties of one and two dimensions (curves and surfaces) could be resolved; a singular variety could be replaced with a similar one that has no singularities. Zariski also showed that, in some cases, singularities in three dimensions (solids) could be resolved, but that proof was restricted to fields of characteristic zero. (A field of characteristic zero consists of any element x such that when x is added to itself, any number of times, the sum is never zero.)

"For many years, this work was considered by everyone in the field to be technically the most difficult proof in all of algebraic geometry," Mumford noted.[45] Teissier, a colleague who spent time at Harvard as a research fellow, compared Zariski's work on singularity resolution to "a cathedral: beauty locally everywhere, directed towards a single global purpose, and a feeling of awe."[46]

Rather than being content to stop at three dimensions, Zariski hoped to construct a general proof, valid for varieties of all dimensions. At a 1954 conference at Princeton honoring Lefschetz's seventieth birthday, Zariski gave a lecture sketching out his approach to proving the resolution of singularities in any dimension. He had not yet completed the proof but believed he was 90 percent finished and felt confident he could carry it through to the end. After the lecture, Arthur P. Mattuck, one of Zariski's Harvard postdocs at the time, asked Lefschetz what he thought of Zariski's work in progress. Lefschetz replied: "Let me tell you something, Mattuck! In the theory of the resolution of singularities, 99 percent equals zero!"[47] By that, Lefschetz meant that a mistake in the last line could topple the entire argument. Unless a proof is perfect, 100 percent correct, it is nothing at all.

And as things developed, Zariski's "90 percent proof" turned out to be flawed. His graduate student Shreeram Abhyankar soon showed that the method would not work. It took a decade, in fact, until another Zariski graduate student, Hironaka, solved the problem using an entirely different strategy. But before that happened, Abhyankar would achieve a breakthrough of his own. He and Hironaka were able to forge ahead of their teacher on this problem, in part, because Zariski did such a good job of paving the way. Abhyankar demonstrated a new approach to the problem, while inventing a battery of techniques to overcome difficulties that had thwarted his predecessors. The methods of modern algebra that Zariski imported brought not only rigor to the problem, Hironaka said, but "clarification" as well.[48] They also brought renewed interest and excitement.

Following Zariski's lead, Abhyankar set out to tackle the problem of resolving singularities of surfaces, solids, and higher-dimensional varieties over a field of characteristic p. Despite his initial enthusiasm, he soon got discouraged, as one of Zariski's suggestions after another led him to a dead end. Zariski himself underwent ulcer surgery during this time (while visiting Italy in 1953), telling his student "that a 3-dimensional singularity was removed from his stomach."[49] At one point, Abhyankar almost quit. Zariski told him not to worry; he would help him find another thesis topic. But rather than give up, Abhyankar worked even harder, including a nonstop stretch of seventy-two hours that yielded his first positive result. His solution to the problem—characteristic p resolution up to dimension 2—was presented in a 1956 paper.[50] Though he hoped to do the same for any dimension, it took him a decade to solve the problem for dimension 3, and the general problem remains unsolved to this day.

Another huge advance on the singularity front came from Hironaka, who met Zariski in 1956 at Kyoto University. Zariski was lecturing there for a month on the subject of algebraic surfaces. He was especially impressed by the work of one of the Kyoto graduate students, Hironaka, who attended all of his lectures. (Hironaka became a mathematician only after his plans for a career as a concert pianist were dashed in junior high school following a botched performance of a Chopin impromptu.) At some point during his visit, Zariski asked Hironaka to discuss his work. Hironaka was so petrified in the presence of Zariski that he "could not even speak the sentences I prepared beforehand," he recounted. "I thought that he [Zariski] was completely disappointed with my poor presentation." But to his surprise, Zariski recommended that Hironaka's work be published in an American journal. He also encouraged him to come to Harvard as a graduate student.[51]

Hironaka took the advice, enrolling in Harvard in 1957. By that time, he was already interested in the resolution of singularities. He had heard about Zariski's work in that area while he was a third-year student in Kyoto and found the problem interesting. He decided early on that he would try it, eventually, even though he was not ready to take it on then. Why did that problem, among all the possibilities, catch his eye? "It's like a boy falling in love with a girl," he said. "It's hard to say why. Afterward you can make up all sorts of reasons." Since Zariski had already solved the lower-dimensional problem, what remained was the problem of higher dimensions. "In the higher dimensions you cannot see everything, so you must have something, some tool, to guess or formulate things," Hironaka says. "And the tool was algebra, unquestionably algebra."[52]

Hironaka's proof relied, to a high degree, on induction. However, when doing a proof by induction, sometimes it is easier to prove a stronger statement than the actual statement you are trying to prove, which in this case concerned the resolution of singularities. The key step Hironaka took was in crafting a stronger statement that encompassed the original problem. He then showed, by induction, that obtaining a result in dimension n would imply, by extension, the same result in dimension $n + 1$. Employing this technique, Hironaka demonstrated that, in the characteristic zero case, singularities could be resolved in *all* dimensions.

He earned a Fields Medal in 1970 for his work, and his 1964 proof is viewed as a tour de force, though it was not universally recognized as such at the time. Attitudes have surely changed, and the resolution of singularities is now accorded more respect than it used to be. Hironaka lectured in Germany shortly after the proof came out, and a professor there said that he had "proved a great useless theorem," Hironaka recalls. "Traditional mathematics used to concern itself with smooth things like balls and donuts. For many, many years people didn't care about singularities. But when the subject matured, they started to realize that singularities are ubiquitous. And new techniques were developed to handle them." For example, the Poincaré conjecture, which constituted one of the most important mathematical problems in the last century, did not say anything about singularities. "But to prove the conjecture, you have to deal with singularities," Hironaka adds.[53]

He became a Harvard professor in 1968 and remains on the faculty in an emeritus capacity. However, back in his graduate school days, Hironaka was in constant contact with two other Zariski students, Artin and Mumford, both of whom have since made outstanding contributions to algebraic geometry. (In addition, Artin served as president of the American Mathematical Society about three decades after their graduate school years, and Mumford won a Fields Medal in 1974.) Hironaka advises students to try to study with "the best scholar in the field. But don't expect you can learn from him! The amazing thing is that with that kind of person, there are many talented young people around, and you can learn a lot from them." He, Artin, and Mumford (along with a younger graduate student, Steven Kleiman, who sometimes joined in) often conducted their own seminars. "We were students, so we had a lot of time to talk about mathematics," Hironaka said. They also attended every lecture Zariski gave. "But he was quite busy as chairman of the department," Hironaka added. "Artin and Mumford and I were always together, always talking, and that was very helpful. So we were influenced not only by our teacher but by our classmates as well."[54]

The three students formed a close bond but almost never collaborated with each other on papers, choosing instead to branch off into separate areas of algebraic geometry. "That was a marvelous period when the three of us were graduate students working with Zariski, who was then in the prime of his career," Mumford reflects.[55] His first encounter with Zariski came about when a fellow student urged him to attend the professor's lecture, "'even though we won't understand a word,' and Oscar Zariski bewitched me. When he spoke the words 'algebraic variety,' there was a certain resonance in his voice that said distinctly that he was looking into a secret garden. I immediately wanted to be able to do this too. It led me to 25 years of struggling to make this world tangible and visible."[56]

During the quarter century or so when Mumford worked in algebraic geometry, which took place roughly between 1958 and 1983, he got his Ph.D. (in 1961) and advanced through the ranks of the Harvard faculty to professor, chaired professor, department chairman (a position that, by tradition, rotates according to seniority), internationally renowned mathematician, and a leader in algebraic geometry like Zariski and Grothendieck before him. In developing geometric invariant theory, which may be considered a part of his broader theory of "moduli," Mumford established a fertile new area in mathematics for which he was awarded his Fields Medal.

Another way of thinking about Mumford's geometric invariant theory is that it provides a foundation, as well as a fundamental tool, for constructing moduli spaces. Mumford was especially interested in the moduli space for algebraic curves of fixed genus—the curves, again, representing solutions to polynomial equations. Although this work is obviously quite technical, the moduli space, he explains, is really "a kind of map, a way of packaging all possible algebraic curves into a single universal object. Just as a normal map is a piece of paper whose points stand for all possible physical locations somewhere, the points of the moduli space stand for curves, one point for each distinct type of curve." The map, in other words, "provides a bird's-eye view of the totality of curves."[57]

A long-term interest of Mumford's has been to understand the global structure of this moduli space, which might also be called a classifying space. "One of the most exciting aspects of math is when such a universal object—mathematicians like to call them 'God-given' objects—turns out not to be simple but to have its own inner nature," he says.[58]

To think about moduli spaces in complicated situations involving curves of higher genus, Mumford showed how one might proceed using geometric invariant theory, which he basically invented, drawing on ideas advanced by the German mathematician David Hilbert about seventy years earlier. One problem that Mumford took up related to the fact that the moduli spaces mathematicians had been working with up to that time—which, again, consisted of curves—were known to be "incomplete." In other words, some curves were missing. The trick was figuring out exactly what had to be added to complete these spaces—just enough so that the space was no longer incomplete but no more than was absolutely needed. Mumford solved this problem with the Belgian mathematician Pierre Deligne (later based at the Institute for Advanced Study), figuring out that what was missing were special types of singular curves called "stable curves," which provide a boundary to the moduli space. Mumford and Deligne also showed how to construct the missing curves.

"Their solution has been central to algebraic geometry ever since," says UCLA mathematician David Gieseker, who got his Ph.D. at Harvard under the supervision of Zariski's student Robin Hartshorne. "Almost everybody who works with moduli spaces relies on the Deligne-Mumford method."[59]

Mumford himself, along with his Harvard colleague Joe Harris, used this technique extensively in proving that moduli spaces for curves of odd genus greater than twenty-three are of "general type," which means they cannot be nicely parameterized or mapped by a complex vector space. Harris and David Eisenbud of the University of California, Berkeley, later extended this result, proving that the moduli spaces for curves of even genus greater than twenty-three are also of general type. (The lower genus case is still not well understood.)

The findings of Mumford, Harris, and Eisenbud show not only that curves of high genus are of general type but also that the moduli spaces that encapsulate these curves are of general type. "This is a weak way of saying the moduli spaces mirror the curves they classify—that the nature of the atlas mimics the land that it represents," Mumford explains.[60] Constructing such spaces and trying to understand their structure, he notes, has become "a big enterprise now."[61]

While Zariski was his main teacher, Mumford learned a lot from Grothendieck as well. "I and my fellow students, Artin and Hironaka, had an ideal training as graduate students at Harvard," he says, in part "because Grothendieck came to visit and teach his own extraordinary new insights in algebraic geometry. He created a synthesis of the older tools of the more intuitive geometric Italian school with the newer French ones of cohomology."[62]

It was in the latter area, cohomology (a concept from topology discussed in Chapter 4), where Artin made a lasting impression. The son of the distinguished mathematician Emil Artin, Michael Artin decided to work with Zariski when he was a Harvard graduate student after taking Zariski's class in commutative algebra. Artin met Grothendieck in 1958, when he came to Harvard, and again during a subsequent visit in 1961. By that time, Artin had heard about Grothendieck's idea for étale cohomology. "When [Grothendieck] arrived I asked him if it was all right if I thought about it, and he said yes," Artin recounted. "And so that was the beginning. He wasn't working on it then—he had the idea but had put it aside. He didn't work on it until I proved the first theorem. He was extremely active, but this may have been the only thing in those years that he really didn't do right away, and it's not clear why."[63]

Artin spent 1963–64 at IHES working with Grothendieck, and visited on several occasions after that, during which they coauthored several publications on étale cohomology. The English translation for the French word *étale* is "slack" (as in slack tide), although Artin is not sure how the nautical terminology relates to the mathematics. He explains the motivation for this idea as follows: "If we're talking about algebraic geometry in complex numbers, cohomology theory was already well established," Artin says. "But if you wanted to do cohomology over a finite field, you couldn't use standard topological methods. There had to be a different approach. This was an attempt to take the notion of cohomology and extend it to other settings."[64]

Weil spurred development in this area, according to Artin. "He showed that this new kind of cohomology could help solve problems over finite fields. In particular, he said you could use it to solve the analogue of the Riemann hypothesis for finite fields, and those instincts were borne out." The Riemann hypothesis, proposed by the German mathematician Bernhard Riemann in 1859, is one of the most famous and important problems in all of mathematics. Riemann contended that the distribution of prime numbers, which does not appear to follow any obvious pattern, is related to a complex function that has since been named the Riemann zeta function. The general hypothesis remains an open problem. But in 1973, Pierre Deligne used étale cohomology to prove the analogue for finite fields. The proof—for which Deligne won a Fields Medal in 1978—vindicated Weil's earlier idea, constituting what Artin called "a crowning achievement" for the notion of étale cohomology.[65]

It is part of the natural order of things that a teacher—especially a great teacher—will eventually be surpassed by his or her most distinguished students. The inevitability of this prospect does not necessarily make it any easier to accept, and this appears to have been the case for Zariski. After Hironaka completed his celebrated proof on the resolution of singularities, Mumford observed, "it was perhaps this time more than any other when Oscar really felt that one of his students had done something that he would have liked to do. But one can see in retrospect that it would have been hard for him to carry out the general case with the tools he had at the time. He had developed half of the abstract tools needed but then stopped short in other directions."[66]

In a similar vein, when Mumford became an assistant professor at Harvard in the early 1960s, "I thought I'd show Oscar how great the new ideas of Grothendieck really were." So Mumford taught a course on these ideas. "I thought that Oscar would be really intrigued, but unfortunately he wasn't," Mumford noted. "He never even came to the class."[67]

That said, Zariski remained active and productive throughout almost his entire life. He served as president of the American Mathematical Society from 1969 to 1970, gaining subsequent recognition for his leadership role in mathematics and many achievements in the field. In 1981, for example, Zariski won the Steele Prize "for his work in algebraic geometry, especially for the many ways he has made fundamental contributions to the algebraic foundations of this subject."[68] He won the Wolf Prize in Mathematics in 1981 for harnessing "the power of modern algebra to serve the needs of algebraic geometry."[69] By then Zariski was starting to suffer from Alzheimer's disease, as well as from tinnitus (or ringing in the ears), which impaired his hearing and eventually led to deafness. His last years, consequently, were quite difficult. He battled periodic bouts of depression, taking refuge in mathematics whenever possible. He died in 1986 at his home in Brookline, Massachusetts.

His former student Joseph Lipman characterized Zariski's time in the United States, since leaving Italy in the 1920s, as "60 years of remarkable mathematical activity lasting well into his eighties—an inspiring counterexample to the dictum that the domain of creative mathematics is reserved for the young."[70]

Referring to Zariski's legacy, his Harvard colleague Barry Mazur said: "Beyond the large theories for which he's famous—the basic lore, the facts of life—there is much more, much deeper thought. People haven't explored these deeper things yet, but they will."[71]

In addition to doing much to rebuild the field of algebraic geometry, Zariski helped transform the Harvard mathematics department. During his time there, the department more than doubled in size, becoming a much more international place, as well as a more active center for research in algebraic geometry and other areas.[72] Zariski contributed greatly to all of this. His presence was felt long after he departed from the stage: mathematics was never the same again, nor was Harvard.

Richard Brauer

Like Zariski, Richard Brauer arrived at Harvard midcareer, with an impressive body of accomplishments behind him, yet he still had a lot more to give. Brauer spent the first three decades of his life in Germany before making his way to the University of Michigan, the University of Toronto, and eventually Harvard. His interest in science in general and mathematics in particular was sparked at an early age by his older brother, Alfred, who went on to become a mathematician of some note. Brauer initially wanted to be an inventor but soon recognized that his "interests were more theoretical than practical." He learned a valuable lesson, nevertheless, through his early experimentation, acquiring "the habit of doing things by myself."[73]

As an undergraduate at the University of Berlin—where Brauer was exposed to the likes of Albert Einstein, Max Planck, and Constantin Carathéodory—he was particularly struck by the lectures of the mathematician Erhard Schmidt, David Hilbert's student and collaborator. "It is not easy to describe their fascination," Brauer wrote. "When Schmidt stood in front of a blackboard, he never used notes and was hardly ever well prepared. He gave the impression of developing the theory right then and there," thus lending an air of immediacy and spontaneity to the proceedings.[74]

In contrast to Schmidt, Issai Schur—who became Brauer's thesis supervisor—delivered more polished lectures, in a rapid-fire fashion that could quickly leave students in the dust. Sometimes Schur told the class about problems he could not solve himself. Brauer and his brother took on one of those problems, solving it in a week. Working independently, their classmate Heinz Hopf did the same. The three students combined their results into a single paper; Brauer thus became the coauthor of his first publication in a mathematics journal. Schur suggested another problem to him, involving representations of groups—the solution to which became part of Brauer's doctoral thesis.

Brauer assumed his first academic post in 1925 at the University of Königsberg—a school that he felt had "fallen into neglect" after the departures of Hilbert and Hermann Minkowski, leaving behind "only second-rate mathematicians."[75] But an effort was under way to restore the university's standing, and Brauer labored hard to that end. In 1931, he published an important paper with Emmy Noether and Helmut Hasse, representing "the climax of a long development in the theory of algebras, which began . . . before the first world war," according to the group theorist Walter Feit.[76]

Fleeing Germany in 1933, Brauer accepted a one-year visiting professorship at the University of Kentucky, arranged through the Emergency Committee in Aid of Displaced Foreign Scholars. Brauer enjoyed his time in Kentucky but was delighted to receive an invitation to spend the 1934–35 academic year at the Institute for Advanced Study as an assistant to the eminent mathematician Hermann Weyl, another former German refugee. Brauer called the appointment the "fulfillment of a dream,"[77] and Weyl considered his interactions with Brauer among the happiest experiences of scientific collaboration in his life.[78]

In the fall of 1935, Brauer became an assistant professor at the University of Toronto, where he spent thirteen productive years. "Today, it is hard to conceive of someone, already 34, with mathematical achievements on the level of Brauer's, being offered, and accepting, an assistant professorship," Feit said.[79] In the context of that era, however, one can imagine that an escapee of Nazi Germany would have appreciated the opportunity to pursue his chosen field, even if the position offered him was not commensurate with his stature and accomplishments. And "accomplish" is something Brauer did a lot of while in Toronto. "He achieved five or six great results during that time, any one of which would have established a person as a first-rank mathematician for the rest of their life," claimed Jonathan Alperin of the University of Chicago.[80]

While in Toronto, Brauer poured himself into the study of finite groups and representations thereof. He would pursue this line of research throughout the rest of his career—and indeed, the rest of his life—including the decades he later spent at Harvard, "producing an extraordinary number of outstanding achievements that fit together into a grand theory," noted Alperin, who called Brauer "a colossus with a focus," a turn of phrase that, quite possibly, has never been used before or since.[81] Given that finite groups provide the subject for which Brauer is best known, it is the main topic in our discussion here, despite his successes in number theory and other areas of mathematics.

A finite group, as the name implies, is a group with a finite number of elements. Brauer was a pioneer in a major effort—ultimately involving upward of one hundred people who turned out some five hundred journal articles totaling about 15,000 pages—to classify finite groups of the type known as "simple."[82] A simple group, by definition, is a group with no "normal" subgroups, except for the so-called trivial subgroups—the identity element and the group itself.

Defining a normal subgroup—a concept originally introduced by group theory pioneer Évariste Galois—can get rather involved. For starters, take a group G and a subgroup of G called H. Then take an element of G called a and multiply a, separately, by every element of the subgroup H. The collection of all these products constitutes a coset, which we will call the left coset, aH. Right cosets, Ha, are similarly defined. The subgroup, H, is called normal if the two kinds of cosets coincide—that is, if $aH = Ha$ for each element a. When this is the case, the cosets form a new group called a quotient group, G/H, whose elements consist of all the cosets aH.

The reason this is important is that a simple group, because it lacks normal subgroups, cannot be broken down into smaller groups via the process of forming quotient groups. In this respect, simple groups are analogous to prime numbers, which are divisible only by themselves and one and cannot be written as a product of smaller factors. Finite simple groups are thus the building blocks of all finite groups, just as prime numbers are the building blocks (or factors) of all positive integers. For this reason, says University of Oregon mathematician Charles W. Curtis, "the classification of finite simple groups provides a foundation for all of finite group theory."[83]

Although there are an infinite number of finite simple groups, the classification problem that Brauer and others took up was to prove that these groups fall into a finite number of "families," as well as to identify all the families. Determining whether a finite group is simple or not turns out to be a relatively easy task when the group is abelian, or commutative. A group is considered abelian if the product of any two of its elements, a and b, is the same, regardless of the order in which they are combined. In other words, *ab* always equals *ba*. This is not the case, however, in a nonabelian, or noncommutative, group. (Some mathematicians, by the way, use the term "simple" strictly for groups that are nonabelian.)

The symmetry group of a square, which again we will call *G*, is an example of a nonabelian group. This group has eight elements: four rotations (one, two, three, or four times to the right, for instance) and four "flips" (about the vertical, horizontal, and two diagonal axes). Doing nothing to the square or, equivalently, rotating it four times to the right (or to the left) leaves the square in its original position, and that symmetry operation is the group's identity element. One can establish, without too much difficulty, that this group is nonabelian: Take a square piece of paper, colored differently on both sides, and number the vertices on each side. If, for example, you rotate the square one time (or 90 degrees) to the right and then flip it over its horizontal axis, the square will end up in a different position than if you flipped it over its horizontal axis first and then rotated once to the right. The order of doing the rotation and flip matters, which means these operations do not always commute.

The four rotations constitute a subgroup, *H*, which happens to be "normal." If we take *a*—an element of the group that is not part of this subgroup (one of the "flips," in other words)—and multiply it by *H*, the four different rotations, the product *aH* will consist of all four kinds of flips. Similarly, if we multiply any one of the rotations by *a*, the product *Ha* will also generate all four possible varieties of flips. Because the left coset, *aH*, equals the right coset, *Ha*, for all possible choices of *a*, we say that the subgroup *H*, consisting of the rotations of a square, is normal. The existence of a normal subgroup that is not trivial means the symmetries of a square do not comprise a simple group.

The number of elements in a group, which is called the group's "order," is important for classification. It can be proved rather easily that an abelian group is simple if its order is a prime number. In this case the group order tells you everything you need to know. The situation regarding nonabelian (noncommutative) groups is much more difficult, and that is the problem Brauer and his contemporaries focused on.

There are only finitely many distinct groups of a given order. Brauer considered the general problem of understanding the simple groups of order *n*. "The strategy," explains Curtis, "is to take the complicated notion of a group and replace it by a number—the order of the group—and see what properties of a group can be deduced from a knowledge of its order."[84]

Perhaps the best way to illustrate this approach is by considering a special case. There is, for example, just one simple group of order 168, whose prime factors are 2, 3, and 7, since $168 = 2 \times 2 \times 2 \times 3 \times 7$. However, there may be more than one simple group with the same order. The order 20,160, for instance, has two simple groups that are distinct from each other—or "nonisomorphic," as it is technically called.[85]

If the order of a group is a prime number, as stated before, the group is automatically simple. If the order has two prime factors, which may or may not repeat—such as 24 ($2 \times 2 \times 2 \times 3$)—the group is never simple, a fact that was established more than a century ago. On the other hand, a group whose order consists of three prime factors may or may not be simple. That said, the order of a simple nonabelian group must have at least three different prime factors.

Brauer embarked on the difficult problem of classifying finite simple groups whose orders contained three different prime factors. In a joint paper, for example, he and the Chinese mathematician Hsio-Fu Tuan proved that a finite simple group of an order divisible by three prime numbers—p, q, r, with p and q appearing to the first power—must be of either order 60 (with the prime factors 2, 3, and 5) or order 168 (with the prime factors 2, 3, and 7).[86]

A 1955 paper, "On Groups of Even Order," which Brauer wrote with his former University of Michigan student Kenneth Fowler, "marked the beginning of a new advance in the theory of finite groups," according to University of Warwick mathematician James Alexander Green.[87] Curtis went even further, noting that the Brauer-Fowler paper "opened a way towards the classification of finite simple groups."[88] The paper also provided a strategy for classifying simple groups that became known as "Brauer's program."[89]

Another big step forward came in 1963 with a 250-page proof published by Feit and John G. Thompson concerning groups of odd order.[90] In discussing their work at the 1970 International Congress of Mathematicians in Nice, France, Feit credited Brauer with taking the first critical step that made their proof possible.[91] In 1972, drawing on Brauer's insights and the Feit-Thompson theorem, Rutgers University mathematician Daniel Gorenstein laid out a sixteen-step program for completing the classification effort and proving the so-called Enormous Theorem, which holds that every finite simple group falls into one of eighteen families or is one of twenty-six "sporadic" groups.

A "family" can be thought of as a collection of groups of different dimensions such as the special orthogonal (SO) group, which includes SO(2), the group of rotations of a circle in a two-dimensional plane; SO(3), the group of rotations of a sphere in three-dimensional space; and SO(n), which includes rotations of higher-dimensional spheres in higher-dimensional space. (A caveat is, perhaps, in order here: while this is a perfectly good example of a family, none of the rotation groups in this case is of the finite variety that Brauer and his peers were interested in classifying.)

Sporadic groups, of which twenty-six have been identified and constructed, do not fit neatly into any of the families. Each of them is unique, and each is generated in its own peculiar way. The smallest sporadic group, which has 7,920 elements, was discovered in the late 1800s by the French mathematician Émile Léonard Mathieu. In 1980, University of Michigan mathematician Robert L. Griess constructed a sporadic group with more than 8×10^{53} elements dubbed the "monster group" because it has the largest order of any of the sporadic groups.[92]

Gorenstein—who did his undergraduate work at Harvard under Saunders Mac Lane and his graduate work under Oscar Zariski—died in 1992, before the program he introduced was entirely wrapped up. The final piece of the puzzle—contained in a paper more than 1,200 pages long—was solved in 2004 by Michael Aschbacher of the California Institute of Technology and Stephen D. Smith of the University of Illinois at Chicago. For his work on this front, which has been called "absolutely monumental," Aschbacher received the 2011 Rolf Schock Prize in Mathematics and the 2012 Wolf Prize in Mathematics.[93]

Although Brauer died several decades earlier, in 1977, after having developed some of the key methodology used to carry this project to the end, his contributions were not forgotten. A 1979 paper by Gorenstein started with a "dedication in memory of Richard Brauer for his pioneering studies of finite simple groups." In his introductory remarks, Gorenstein wrote, "It is indeed unfortunate that Richard Brauer did not live to see the complete classification of finite simple groups. He had devoted the past thirty years largely to their study, and it is difficult to overestimate the impact he made on the subject."[94]

While Brauer never saw the program he was so invested in finally reach fruition, Feit said, "he was fortunate in that he lived long enough to receive the recognition which he deserved. He was even more fortunate in that his interest and abilities lasted to the end."[95]

"It is a striking fact of [Brauer's] career that he continued to produce original and deep research at a practically constant rate until the end of his life," Green noted. "About half of the 127 publications which he has left were written after he was fifty."[96] Given that Brauer was fifty-one when he came to Harvard, the university's investment in him was amply rewarded.

Raoul Bott

The hiring of Raoul Bott also paid off immensely for Harvard, although it is fair to say that as a young man, Bott did not exhibit great mathematical flair, nor did he show much academic promise in general. Born in Budapest in 1923—and raised mainly in Slovakia (until his family immigrated to Canada in 1938)—Bott was, at best, a mediocre student throughout childhood. In five years of schooling in Bratislava, Slovakia, he did not earn a single A, except in singing and German. In mathematics, he typically got Cs and the occasional B, which should make him a hero among late bloomers.

As a youth of about twelve to fourteen, Bott and a friend had fun playing around with electricity—creating sparks, wiring together fuse boxes, transformers, and vacuum tubes, and, in the process, figuring out how various gadgets work. This experimentation eventually served him well. A mathematician, Bott later explained, is "someone who likes to get to the root of things."[97]

Although Bott frequently told his Harvard students that he never would have made it into the school as an undergraduate, he somehow managed to get into McGill University, where he majored in electrical engineering.[98] Upon graduating in 1945, he joined the Canadian army but left after four months when World War II came to a sudden and unexpected end.

Bott then enrolled in a one-year master's program in engineering at McGill, but after completing that he was still unsure of his future course. So he met with the dean of McGill's medical school, as he contemplated a career shift. In response to questions posed by the dean, Bott confessed to disliking the dissection of animals and hating chemistry even more, showing little enthusiasm for the standard subjects taken up in medical school. Finally, the dean asked him: "Is it that maybe you want to do good for humanity...? Because they make the worst doctors."[99]

That ended Bott's thoughts of a medical career. "I thanked him," Bott said. "And as I walked out of his door I knew that I would start afresh and with God's grace try and become a mathematician."[100]

Initially, he wanted to pursue mathematics at McGill, until he was told that his background was so thin he would have to get a bachelor's degree first—a process that could take three years. He turned instead to the Carnegie Institute of Technology (since renamed Carnegie Mellon University), interested in the master's program in applied mathematics. But the course requirements were so extensive that it would have taken him three years to get a master's degree. Carnegie's math chair, John Lighton Synge, suggested their new doctoral program, which had hardly any requirements at all. Bott liked the idea and was assigned Richard J. Duffin as his adviser.

Bott and Duffin took on, and eventually solved, what was then one of the most challenging problems in electrical network theory. The resultant Bott-Duffin theorem not only was of great theoretical significance but also had important practical applications in the electronics industry. The paper that Bott and Duffin coauthored had important consequences for Bott's career, as well, because the work impressed Hermann Weyl, who arranged for Bott to spend the 1949–50 academic year at the Institute for Advanced Study in Princeton.[101] (Weyl, as you may recall, came up earlier in this chapter, having secured a position at the Institute for Advanced Study for Richard Brauer; Weyl later did the same for Richard's brother, Alfred.)

"The general plan of my appointment [at the institute], as I understood it," Bott wrote, "was that I was to write a book on network theory at the Institute." On his first day at work in Princeton, Bott met with Marston Morse, who was in charge of the temporary members that year. "[Morse] immediately dismissed my fears of having to write a book. It was a matter of course to him that at the Institute a young man should only do what he wanted to do; that was the place where a young man should find himself and the last place in the world for performing a chore... I remember leaving this interview with a light heart, newly liberated, and buoyed by the energy and optimism I had just encountered."[102]

Bott immediately went to his office and started working on the four-color problem, thinking that the trick he had used to solve the network problem—a function that he considered his "secret weapon," might crack this problem as well. Morse dropped by for a chat a few weeks later. "When he heard what I was doing, he didn't really object," Bott wrote. "Instead, he first spoke of the great interest in the question, but then started to talk about the many good men he had seen start to work on it, never to reappear again. After he left, I threw *all* my computations in the waste basket and never thought about the question again!"[103]

What he did think about—perhaps owing to the presence of Morse, Norman Steenrod, and others—was topology. Surrounded by giants in the field, Bott studied the subject deeply, though he appeared to be in no great rush to publish anything. In fact, he said, "I didn't write a single paper in my first year there. So I was very delighted when Marston Morse called me up at the end of that year and said, 'Do you want to stay another year?' And I said, 'Of course, yes!' He said, 'Is your salary enough?' (It was $300 a month.) I said, 'Certainly!' because I was so delighted to be able to stay another year. My wife took a dimmer view, but we managed."[104]

In 1951, Bott joined the faculty of the University of Michigan, where he continued to focus on topology, paying particular attention to Morse's theory of critical points. A standard picture from Morse theory, as discussed in Chapter 4, involves a doughnut (or "torus") standing upright. This surface has four "critical points"—a maximum on top of the doughnut, a minimum on the bottom, and two saddle points on the top and bottom of the doughnut's inner ring. "Generally, the critical points of a function are isolated," Bott wrote, but he realized these points could come in "bigger aggregates" and could even be the special kinds of spaces we call manifolds.[105]

One way to picture this is to take the upright doughnut from the previous example and topple it over, so that it is now lying on its side, flat on a tabletop. The maximum of this newly configured object is no longer a point—it is a circle. The minimum, similarly, is a circle too. One could determine the topology of the space—and correctly identify it as a torus—by knowing that the critical manifolds are two circles, aligned one above the other.

Bott thus provided a generalization of classical Morse theory, often called Morse-Bott theory, in which critical manifolds replaced the critical points of the original theory. The critical manifolds of this theory could be individual points, which is the zero-dimensional special case. Or they could be one-dimensional manifolds, like the circles in the tipped doughnut. They could be higher-dimensional objects, too—manifolds of any finite dimension, in fact.

Bott used this generalized version of Morse theory to compute the homotopy groups of a manifold, and from there he proved the periodicity theorem. That, admittedly, is a bit of a mouthful, so we will try to break down that statement, explaining in simplified terms what he did.

Whereas Morse was primarily interested in using topology to solve problems in analysis—to solve differential equations, in other words—Bott turned that around, using Morse theory to solve problems in topology. And one of the main problems in topology—as in other areas of mathematics and throughout science, in general—is the classification problem. "Just as scientists want to classify plants and animals to understand how biology works and how life is organized, mathematicians also strive to find some order among mathematical objects," says Tufts University mathematician Loring Tu, a former Harvard graduate student who coauthored a book on algebraic topology with Bott.[106] Group theorists, therefore, are interested in classifying groups, such as the finite simple groups discussed earlier in this chapter. Topologists, similarly, want to be able to look at various spaces—seeing which ones are equivalent, which ones are different—and then sort them into their proper bins.

Mathematicians define invariants—fixed, intrinsic features of a space—in order to distinguish among different topological spaces. If two spaces are "homeomorphic"—meaning that one can be deformed into the other by stretching, bending, or squishing but not cutting—they must have the same topological invariants. One of the simplest topological invariants to define is the homotopy group, there being one for each dimension. If two spaces (or manifolds) have different homotopy groups, they really are different and cannot be homeomorphic.

Computing the homotopy groups of a manifold constitutes an important step toward understanding the topology of that manifold. The first homotopy group, also called the fundamental group, relates to the kinds of loops you can draw in a space that cannot be shrunk down to a point. A two-dimensional sphere, for example, has a trivial fundamental group, because any loop you can draw on the surface of a sphere can be shrunk down to a point without impediment. A sphere's first homotopy group is therefore zero. On a doughnut, there are two different kinds of circles that cannot be constricted to a point. One would be a circle that starts on the outside of the doughnut, goes through the hole in the center, and loops around. This cannot be shrunk to a point without cutting the doughnut. There is also another kind of circle that wraps around the circumference of the doughnut, sticking to what might be called the "equator." It, too, cannot be shrunk to a point without crushing the doughnut so that it no longer has a hole and, therefore, is no longer a doughnut—just some amorphous, mashed-up pastry.

The fundamental group of the doughnut thus has "two generators," two distinct circles, explains Tu, "but you can go around a circle any number of times, in a positive or negative direction, so we say the fundamental group of the doughnut is two copies of the integers." Tu adds that "the homotopy groups are very easy to define, but they are very difficult to compute, even for a two-dimensional sphere, which seems like a simple enough object."[107]

One puzzling feature was that the homotopy groups appeared to follow no pattern whatsoever. The first homotopy group for the sphere (or the fundamental group) is zero, as mentioned before. The second homotopy group contains all the integers, and the third homotopy group contains all the integers, too, whereas the fourth and fifth groups have just two elements, and the sixth group has twelve elements. There was no apparent rhyme or reason to it.

This perplexing situation made Bott curious enough to try computing some of the homotopy groups on his own. Among other things, he was interested in determining the homotopy groups associated with a rotation group of arbitrary dimension, or $SO(n)$, as it is called. Keep in mind, however, that these rotation groups are Lie groups (as discussed in Chapter 6), which means that they are also manifolds. A manifold, in turn, has homotopy groups. So if you want to understand the structure of rotations in n-dimensional space, one of the first things you might try is computing the homotopy groups.

That is what Bott set out to do. Applying Morse theory to study the homotopy groups of Lie groups, he uncovered an astonishing pattern: the "stable" homotopy groups of $SO(n)$—which is to say, the homotopy groups of $SO(n)$ when n is sufficiently large—literally repeat in cycles of eight. For large values of n—or for stable homotopy groups, in other words—the first homotopy group for $SO(n)$ is the same as the ninth homotopy group, and the second homotopy group is the same as the tenth homotopy group, and so on. The same thing happens when one looks at rotations in "complex space," a space with complex number coordinates. These rotations form a group called $SU(n)$, which stands for "special unitary group." When *n*, once again, is sufficiently large, the homotopy groups of SU(*n*) repeat in cycles of two: the first homotopy group for SU(*n*) is the same as the third homotopy group, the second homotopy group is the same as the fourth, and so on.

Bott's 1957 paper—which established this finding and was expanded upon in later work—came as a "bombshell," according Michael Atiyah, a longtime collaborator of Bott's presently based at the University of Edinburgh. "The results were beautiful, far-reaching and totally unexpected."[108]

Some mathematicians have compared the periodicity theorem to the periodic table of elements in chemistry. Hans Samelson, a University of Michigan colleague whom Bott considered a "kindred spirit,"[109] called the "periodicity result . . . the loveliest fact in all topology, with its endlessly repeatable 'mantras' . . . The discovery had a tremendous effect and started a flood of developments."[110]

Some of the developments Samelson alluded to included K-theory, the study of vector bundles that was pioneered by Grothendieck, Serre, Atiyah, and Friedrich Hirzebruch. (A cylinder is a simple example of a vector bundle consisting of vectors—in this case, vertical line segments or "arrows" endowed with both a direction and magnitude—attached to a circle lying in a horizontal plane.) A 1959 paper by Bott provided a "K-theoretic formulation of the periodicity theorem," and several years later he and Atiyah provided a new proof of periodicity, which fit into the K-theory framework.[111] The periodicity theorem was extremely useful in this context, because it provided an expeditious way for mathematicians to classify vector bundles. As Harvard mathematician Michael Hopkins explains, "The periodicity theorem accounts for the computability of K-theory (and in some sense K-theory itself)."[112] Consequently, Bott added, "K-theory then took off, and it was great fun to be involved in its development."[113]

Thanks to his proof of the periodicity theorem, Bott received offers from four universities and did not know what to do. Harvard mathematician John Tate urged his school—which had no topologist at the time—to hire Bott. Zariski, who was the mathematics chair at the time, liked the idea, figuring that "Bott was just the man to enliven what often seemed to him a rather stodgy department."[114] Bott accepted the offer in 1959 and stayed for the rest of his career.

Five years later, at a conference in Woods Hole, Massachusetts, in 1964, Atiyah and Bott came up with a formula that Bott considered "among my favorites in all of mathematics."[115] Their work was a broad generalization of the Lefschetz fixed-point theorem, which Princeton mathematician Solomon Lefschetz proved in 1937. Lefschetz's formula (which is expressed in terms of cohomology) involves the number of fixed points of a map from a space to itself. A map, as stated previously, is like a function that takes a point in one space and assigns it to a point in another space (although the "other" space could, in fact, be the same space).

As a simple example, suppose we have the function $g(x) = 3x^4 + 2x + 1$ and want to solve the equation $3x^4 + 2x + 1 = 0$. One thing we can do is add an x to both sides of the equation: $g(x) + x = x$. Next, we invent a map $h(x)$ that is equal to $g(x) + x$. Then the original equation $g(x) = 0$ is equivalent to $h(x) = x$, which means that h maps x to itself. Thus, a solution x of the original equation is a fixed point of the map h. It is called a fixed point because x does not change during the mapping from one space to another; its position remains fixed. (Technically speaking, this example concerns an algebraic equation on the real line; the same idea of transforming a solution of an equation to a fixed point of a map applies to a differential equation on a manifold.)

To see the Lefschetz fixed-point theorem in action, rotate a sphere around its vertical axis. That is an example of a transformation, or map, that takes a sphere to a sphere. In this case, there are just two fixed points—the north and south poles—as every other point moves during the rotation. One can also use Lefschetz's formula to determine algebraically that there are two fixed points. A virtue of the latter approach is that you can use it to figure out the number of fixed points in more complicated situations where you cannot draw a simple picture of a spinning globe.

"Algebra is almost always easier to do than geometry and topology, and that's the basic idea behind cohomology, which involves converting geometric and topological problems into algebraic ones," explains Tu. Atiyah and Bott went further still, Tu adds, "deriving a far-reaching generalization that, in one special case, gives you back the Lefschetz theorem, but it also gives you many other fixed-point theorems—some new theorems and some classical ones, as well."[116]

In 1982, Atiyah and Bott came up with another formula involving fixed points, this one on the subject of "equivariant cohomology." Cohomology, a term that comes up many times throughout this book, is an algebraic invariant that mathematicians assign to spaces, which means it is one of the tools they use to study spaces. When the space you are studying has symmetries of a particular sort, there is a kind of cohomology you can study called "equivariant cohomology."

Two French mathematicians, Nicole Berline and Michèle Vergne, discovered this same formula independently, and almost simultaneously. The Atiyah-Bott-Berline-Vergne formula—or the equivariant localization formula, as it is often called—allows you to compute certain integrals on manifolds with symmetries. This formula is extremely convenient since many important physical quantities can be expressed as integrals, yet computing those integrals can be quite difficult.

Consider again the example of a sphere, this time of radius 1. It has rotational symmetry about a vertical axis and, as before, has exactly two fixed points. The surface area of the sphere is a surface integral. The equivariant localization formula assigns a number, or multiplicity, to each fixed point and says that the surface integral is a constant, 2π, times the sum of the multiplicities at all the fixed points. In this case, the multiplicity assigned to each fixed point is 1, and since there are two fixed points, the area of the unit sphere, according to this reckoning, is 2π times 2, or 4π, as it should be. The approach is "very powerful," says Tu, "because instead of having to compute an integral, you just have to add up a few numbers."[117]

Commenting on their eventual breakthrough, Bott said that "Michael and I had been wrestling with the question of equivariant cohomology since the 1960s," approaching it through the notion of fixed-point theorems. "It is amazing how long it takes for a new idea to penetrate our collective consciousness and how natural and obvious that same idea seems the moment it is properly enunciated."[118]

Speaking of his long-term collaboration with Bott, whom he knew for more than fifty years, Atiyah said: "It was impossible to work with Bott without becoming entranced by his personality. Work became a joy to be shared rather than a burden to bear . . . His personality overflowed into his work, into his relations with collaborators and students, into his lecturing style, and into his writing. Man and mathematician were happily fused."[119]

Another colleague, the Harvard geometer Clifford Taubes, maintains that Bott had a profound influence on him when he was a graduate student at Harvard, earning his Ph.D. in physics in 1980. "It was just wonderful . . . to see how this beautiful mathematics flowed," Taubes said of the class he took that was taught by Bott. "I, for one, would have been a physicist if I had not been in his class, but I was seduced by mathematics." Bott's impact, of course, spread far beyond a single graduate student in the late 1970s. All told, says Taubes, "he had a tremendous influence in the development of modern geometry and topology. I would say that his contributions to this were as great as any one person."[120]

Taubes is not alone in that assessment. Bott won the 2000 Wolf Prize in Mathematics, which he shared with Jean-Pierre Serre, for his work in topology that culminated in the periodicity theorem and "provided the foundation for K-theory, to which Bott also contributed greatly."[121]

Bott died in 2005, after having proved many important theorems and leaving an indelible mark on generations of students. Two of his students won the Fields Medal: Stephen Smale, who got his Ph.D. at the University of Michigan in 1957, and Daniel Quillen, who got his Ph.D. from Harvard in 1964. Another of his Harvard students, Robert D. MacPherson—a coinventor of "intersection homology"—has had a distinguished career at Brown University, MIT, and the Institute for Advanced Study.

Bott was, by all accounts, an imperturbable teacher. Once a five-square-foot chunk of ceiling fell down in the middle of his Math 11 classroom. He calmly waited for the dust to settle and then resumed his discussion, urging his students to ignore the large cracks in the ceiling.[122] A class with Bott, says Benedict Gross, "was an amazing experience, like drinking from the original stream, as a lot of it was his own work."[123]

"I recall him arriving at each class with no notes, puffing on a cigarette right under the No Smoking sign, and simply living the mathematics in our presence," said Washington University mathematician Lawrence Conlon, who was a graduate student when Bott burst onto the Harvard scene. After a mind-blowing class with Bott on algebraic topology, Conlon mustered the courage to ask him to direct his thesis. "Well, Larry," said Bott, "you're a good student, but what we have to find out now is whether you can dream."[124]

Bott's presence, according to Mazur, "radiated friendship of the sort that simply made everyone not only happier but somehow perform better."[125] Bott was that rare person, possessing "such an extraordinary amount of humor and optimism" that he could truthfully claim, as he did, "I can't say that there is any mathematics that I don't like."[126]

Much as his students and colleagues appreciated Bott, he also appreciated them in return. Upon receiving the Steele Prize for Lifetime Achievement in 1990, he offered thanks for what had then been more than thirty years at Harvard, where "there is not a single colleague or student who has not added to my education or uncovered some hidden mystery of our subject."[127]

In hiring Ahlfors, Zariski, Brauer, Bott, and others who followed, Harvard was opening its doors to mathematicians from foreign shores who enriched the department, the field, and the culture. Bott extended thanks in return to "this country, which has accepted so many of us from so many shores with such greatness of spirit and generosity. Accepted us—accent and all—to do the best we can in our craft as we saw fit."[128]

EPILOGUE

Numbers and Beyond

While trying to lure Raoul Bott to Harvard in 1959, John Torrence Tate was facing a quandary of his own. Although he had been employed at Harvard since 1954, Princeton had just made him an offer—sparked by some exciting work Tate had just completed while on sabbatical in Paris—that was too tempting to ignore. Bott's decision to come to Cambridge helped persuade Tate to stay put; the presence of the talented, as well as charismatic, Bott in the department instantly made Harvard a more attractive place to be. Tate's decision, in turn, had profound consequences for Harvard, as it led to the establishment of a vibrant number theory group at the school that remains strong and growing.

Throughout his lengthy career, Tate has made vital contributions to so many areas of number theory that it is hard to single out just one. A large and important body of his work concerns elliptic equations (of the form $y^2 = x^3 + ax + b$) and the curves they define, as well as the higher-dimensional versions of elliptic curves called abelian varieties. Elliptic curves, as the above equation indicates, are described by cubic polynomials in two variables. They are associated with so-called elliptic integrals, which can be used to calculate the arc lengths of ellipses. They are also, as Tate puts it, the first nontrivial examples of abelian varieties.

Although the equations that describe elliptic curves may look simple at first glance, they are extremely challenging to solve. And the theory of elliptic curves is surprisingly deep, interacting with many distinct branches of mathematics. Tate wrote a number of hugely important papers that illustrate how these curves work, while clarifying their uses in mathematics. He also coauthored a book on the subject—with his former Harvard Ph.D. student and current Brown professor Joseph Silverman—that was based on lectures Tate had delivered at Haverford

College in 1961. "John Tate offered a new way of looking at elliptic curves that transformed our whole perspective," notes Harvard mathematician Benedict Gross, another former student of Tate's.[1]

By greatly enhancing our understanding of elliptic curves, Tate helped advance many important problems in mathematics. The Birch and Swinnerton-Dyer conjecture, which has inspired much exciting work, addresses the question of whether elliptic equations have a finite or infinite number of rational solutions. The conjecture, which was formulated in the mid-1960s, is still an open problem, having been proven only in a few special cases. Yet Tate has done much to refine the conjecture and frame it in a way that is more readily approachable. Elliptic curves, moreover, were central to the proof of Fermat's last theorem in 1995 by Andrew Wiles, a Princeton University mathematician who spent three years (1977–1980) as a Benjamin Peirce Instructor at Harvard. Wiles proved the general theorem, but special cases had been proven much earlier by Pierre de Fermat himself, Leonhard Euler, and others.

A 1978 paper by Harvard mathematician Barry Mazur about the rational points of elliptic curves contributed to important advances in number theory and arithmetic algebraic geometry, including Wiles's proof of Fermat's last theorem.[2] Mazur first came to Harvard in 1959 as a junior fellow in Harvard's Society of Fellows, after getting a Ph.D. from Princeton and spending a year at the Institute for Advanced Study. By that time, Mazur had already made a name for himself by proving the generalized Schoenflies problem, the two-dimensional version of which argues that any simple closed curve in the plane encloses a region that is topologically equivalent, or "homeomorphic," to a disk.

Mazur joined the Harvard faculty in 1962, which was around the time he made the switch from topology to number theory. His conversion came about somewhat by accident, while he was exploring knot theory, a branch of topology. "I knew a lot about knot theory and came across an analogy between knots and prime numbers," Mazur recalls. "If you really push this analogy far, knot theory can take you close to something deep in number theory."[3] He followed this thread, taking it as far as he could go and, for the most part, never came back. That may explain reports of a story, which allegedly circulated, "about this great young topologist who disappeared off the face of the earth," only to emerge, unbeknownst to the tellers of that tale, in the mysterious world of number theory.[4]

Together, Tate and Mazur formed the nucleus of a robust number theory group at Harvard, which was enlivened by periodic visits from

Jean-Pierre Serre, Alexander Grothendieck, Serge Lang, and many others. While there had been little research along these lines at Harvard prior to Tate's arrival, things started reaching a critical mass with both him and Mazur on the scene. During the 1960s, Mazur recounts, "number theory was emerging at Harvard in a big way, and it was becoming an exciting place." During the latter part of the decade, he and Tate started a number theory seminar, which has been going practically every week for more than forty-five years. "John and I saw it as a general forum for finding out what was new and interesting in number theory, mainly drawing on local talent. We had no budget and would rarely invite people who would have to take an airplane to get here. It was also an informal thing—a person would get up and talk with minimal introduction. We did it that way because it suited our temperaments and because we thought it would be sustainable in the long run, although we never expected it to run this long."[5] Over the decades, the seminars became bigger, drawing in more people who came from afar, including those who traveled by air.

Speaking of travel, Tate took a rather pronounced hiatus from Harvard in 1990, when he accepted a chaired position at the University of Texas at Austin. After retiring from Texas in 2009, he returned to Harvard in an emeritus capacity. Now in his late eighties, Tate still attends the number theory seminars that he and Mazur started so many decades ago.

One thing that he probably cannot help but notice is how often his name comes up in discussions of number theory. His ubiquity in the field is demonstrated, in part, by the number of important terms that bear his name: the Tate module, Tate motive, Tate curve, Tate cycle, Tate cohomology, Tate's algorithm, Hodge-Tate decompositions, Serre-Tate deformation theory, Lubin-Tate group, Tate trace, Tate-Shafarevich group, Néron-Tate height, Sato-Tate conjecture, Honda-Tate theorem, and so on.

For Tate, this seems to be more a source of embarrassment than pride. "I never sought to have my name attached to all those things," he says. "It's mainly a consequence of the fact that I did not publish a lot, so that other people [especially Serge Lang] ended up naming them. It may also stem from the fact that 'Tate' is such a nice, short name." As for why he publishes less than he could, and less than other people in the field might desire, some colleagues believe that Tate is a bit of a perfectionist (and perhaps more than a bit) when it comes to his own writing. "I tend to write something up and then tear it up and start over, always trying to make things better," Tate admits. "Eventually I find reasons not to publish it."[6]

Fortunately, Tate is more comfortable writing letters, and many of his ideas got into circulation that way. "He'd write letters to people like Serre and [the British number theorist John] Cassels with amazing mathematical ideas in them," Silverman says. "For example, there's something called 'Tate's algorithm' for finding the minimal model of an elliptic curve, which he sent to Cassels. Eventually Tate agreed to let Cassels publish the letter, verbatim, in a conference proceedings. And I wrote a paper that's an improvement of another algorithm of Tate's that had previously only appeared in a letter he wrote to Serre."[7]

The good news is that a large number of Tate's ideas did get out into the broader community, and many of them have been extremely influential. Given his initial hesitance to enter the field—he started graduate school at Princeton enrolled as a physics student, afraid he would not measure up as a mathematician—Tate confesses to being "surprised that I was actually able to accomplish things in mathematics, some good, no doubt."[8]

He attributes his success, to the extent he will admit to any, to a couple of factors. One is that he has never been diverted by a serious hobby and has instead kept his focus on mathematics, perhaps to a greater degree than some of his more rounded peers. "I'm certainly not a Renaissance man. I don't have wide knowledge or interests," Tate admits. "My feeling is that to do some mathematics, I just have to concentrate. I don't have the kind of mind that absorbs things very easily." Another thing that has helped him is the fact that you do not have to be a speedy problem solver to get anywhere in mathematics. "It doesn't matter how long it takes," he says, "if the end result is a good theorem."[9]

A "good theorem," as Tate puts it, lasts forever. Once proved, it will *always* stay proved, and other mathematicians are free to use it and build on it as they please, sometimes to great effect. In addition to proving theorems, mathematicians can contribute to their field in other ways, by asking stimulating questions, for example, or by inspiring students and spurring them on to achievements of their own. Tate and his colleague Raoul Bott did all of the above and more. But they also made a contribution in another area as well: they helped secure a new home for Harvard's math department so that all the faculty, students, and staff could finally occupy the same space, with an architecture that promoted accessibility and cohesiveness, while facilitating interactions among the occupants.

"When I was an undergraduate here in the 1940s, there was no math building and no discernible department in terms of something you could see," Tate says. "Professors had offices all over the place, and some might have had no offices at all." Most classes at the time were taught in Sever Hall, which had a small room for professors to hang their coats in before class. "That was as much of a common area as we had."[10]

After World War II, most math offices and staff were moved temporarily to a Quonset hut until they moved again, around 1950, to the old geography building at 2 Divinity Avenue. But that setup was hardly ideal because the only potential classroom was a lecture hall on the first floor. Adding to the inconvenience, teaching assistants were stationed on Cambridge Street a couple of blocks away.[11]

When the new (and current) mathematics quarters in Harvard's Science Center were being readied in the late 1960s and early 1970s, Bott got involved in a big way. With the help of Tate, who was then the department chair, Bott fought to have a common room that was literally in the middle of things, located on the fourth floor when the department was to occupy the third, fourth, and fifth floors. (The math department has since expanded to include the first and second floors as well.) Bott wanted these floors to be interconnected by multiple stairways, and he also pushed for an open staircase leading to the common room.

"Raoul attended meetings with the architects because he believed that the physical space for the department would affect its atmosphere of collegiality," Mazur says. "He was absolutely insistent that there be windows that could be opened, which made things more complicated for the design of the heating and cooling system." He was also unyielding about the need for an open stairway, which made it more difficult to meet the fire codes. He approached these challenges with the same ingenuity that he applied to problems in math. He usually prevailed in math, and with the backing of Tate and others, Bott prevailed in these matters as well. He could be, as those who knew him affirm, *very* persuasive. Fortunately, Mazur adds, "the effort paid off, as the building has been surprisingly successful."[12] Tate agrees: "It's good to have everyone under the same roof. It really does make a difference."[13]

Looking back at the move to the Science Center, some forty years later, it is fair to say that the department has thrived in its current accommodations. Although this book focuses mainly on the years 1825–1975, we are happy to report that math is alive and well at Harvard to this day. At a luncheon honoring Tate after he won the 2010 Abel Prize, Tate shared some of the glory with the department itself, contending that it was still "as strong as ever." In fact, he believes the department is even stronger today because it has grown in so many different areas since he first became associated with Harvard, when he arrived as an undergraduate more than seventy years ago.[14]

Shlomo Sternberg, a versatile mathematician who has been at Harvard for more than half a century, is "very optimistic about the future." Perhaps best known for his work on dynamical systems and symplectic geometry—a branch of differential geometry and topology that focuses on even-dimensional spaces—Sternberg joined the faculty in 1959, the same year that Bott and Mazur came to Harvard. All three were relatively young men at the time. And thanks to much more recent appointments of youthful mathematicians, including the hiring of Dennis Gaitsgory, Jacob Lurie, and Mark Kisin, "our department is not overloaded with old people," Sternberg maintains.[15]

Young scholars are the key not only to the future of the department but also to the field itself, adds Benedict Gross. "They're the ones who bring the really great ideas into the world. We've made six appointments in the past five or six years, almost all of them people in their 20s when we appointed them. They invigorate the department."[16]

With these newcomers working alongside the veteran players, pushing forward on some familiar fronts while opening new frontiers on others, it is clear that the history of mathematics is still being written at Harvard. The story continues to unfold, with today's researchers building on progress in the field that has been under way for centuries and, indeed, for millennia. "Sometimes a line of mathematical research extending through decades can be thought of as one long conversation in which many mathematicians take part," notes Mazur, who, like Sternberg, has been on the faculty for more than fifty years and remains an active participant in that "conversation."[17]

In the preface of this book, we compare mathematics to a river that flows through time, space, and the human mind. Looking at Harvard's math department today, and at the field as a whole, one can confidently say that the river is going strong, sprouting new tributaries as we speak. But unlike normal rivers, whose origins and endpoints are pretty well fixed, we do not know exactly where this river is going. We cannot control its direction, nor should we try. About the only thing we can say for sure is that it will undoubtedly take us places we have never been before, leading to untold adventures ahead.

NOTES

Prologue

1. "The John Harvard Window," *The Harvard Graduates' Magazine* **14** (September 1905), pp. 198–200.

2. Edmund B. Games Jr., "The Start of Harvard Education," *Harvard Crimson*, June 12, 1958, available at http://www.thecrimson.com/article/1958/6/12/the-start-of-harvard-education-pbwhen/.

3. Samuel Eliot Morison, *Three Centuries of Harvard: 1636–1936* (Cambridge, Mass.: Harvard University Press, 1946), pp. 9–10.

4. Ibid., p. 30.

5. Florian Cajori, "The Teaching and History of Mathematics in the United States," Bureau of Education, Circular No. 3, 1890, pp. 18–28.

6. Samuel Eliot Morison, *Harvard College in the Seventeenth Century* (Cambridge, Mass.: Harvard University Press, 1936), p. 208.

7. Samuel Eliot Morison, *The Founding of Harvard* (Cambridge, Mass.: Harvard University Press, 1995), p. 435.

8. Cajori, "Teaching and History of Mathematics."

9. Ibid.

10. Morison, *Harvard College,* pp. 208–209.

11. Dirk J. Struik, "Mathematics in Colonial America," in *The Bicentennial Tribute to American Mathematics,* edited by Dalton Tarwater (Washington, D.C.: Mathematical Association of America, 1977), p. 3.

12. Benjamin Peirce, *A History of Harvard University* (Cambridge, Mass.: Brown, Shattuck, 1833), p. 187.

13. Morison, *Three Centuries of Harvard,* p. 92.

14. J. L. Coolidge, "Three Hundred Years of Mathematics at Harvard," *American Mathematical Monthly* **56** (1943), pp. 349.

15. Cajori, "Teaching and History of Mathematics."

16. Coolidge, "Three Hundred Years of Mathematics," p. 349.

17. Morison, *Three Centuries of Harvard*, pp. 190, 195.

18. Bryan Norcross, *Hurricane Almanac: The Essential Guide to Storms Past, Present, and Future* (New York: St. Martin's Press, 2007), p. 96.

19. Peirce, *A History of Harvard University.*

1. Benjamin Peirce and the Science of "Necessary Conclusions"

1. Florian Cajori, "Teaching and History of Mathematics in the United States," Bureau of Education, Circular No. 3, 1890, pp. 18–28.

2. Lao Genevra Simons, *Fabre and Mathematics* (New York: Scripta Mathematica, 1939), p. 49.

3. F. P. Matz, "Benjamin Peirce," *American Mathematical Monthly* II (June 1985), p. 174.

4. Raymond Clare Archibald, *A Semicentennial History of the American Mathematical Society, 1888–1938* (New York: American Mathematical Society, 1938), p. 1.

5. Somerville's work in this area and her book were brought to the authors' attention by Sir Michael Atiyah of the University of Edinburgh.

6. Joseph Lovering, "The Mécanique Céleste of Laplace, and Its Translation with a Commentary by Bowditch," in *Proceedings of the American Academy of Arts and Sciences: From May 1888 to May 1889* (Boston: John Wilson and Son, 1889), pp. 185–201.

7. Sven R. Peterson, "Benjamin Peirce: Mathematician and Philosopher," *Journal of the History of Ideas* 16 (January 1955), p. 95.

8. Samuel Eliot Morison, *Three Centuries of Harvard* (Cambridge, Mass.: Harvard University Press, 1965), p. 264.

9. Benjamin Peirce, *A System of Analytic Mechanics* (Boston: Little Brown and Company, 1855), p. v.

10. Cajori, "Teaching and History of Mathematics," p. 133.

11. R. C. Archibald, "Benjamin Peirce. Biographical Sketch." *American Mathematical Monthly* 32 (January 1925), pp. 8–19.

12. Benjamin Peirce, "On Perfect Numbers," *Mathematical Diary* 2 (1832), pp. 267–277.

13. Edward J. Hogan, *Of the Human Heart* (Bethlehem, Penn.: Lehigh University Press, 2008), pp. 98–99.

14. Benjamin Peirce, "On Perfect Numbers," *Mathematical Diary* 2, p. 267.

15. Jennifer T. Betcher and John H. Jaroma, "An Extension of the Results of Servais and Cramer on Odd Perfect and Odd Multiply Perfect Numbers," *American Mathematical Monthly* 110 (January 2003), p. 49.

16. Pace P. Nielsen, "Odd Perfect Numbers Have at Least Nine Distinct Prime Factors," *Mathematics of Computation* 76 (2007), pp. 2109–2126.

17. Hogan, *Of the Human Heart*, p. 69.

18. "Sketch of Professor Benjamin Peirce," *Popular Science Monthly* (March 1881), pp. 691–695, HUG 300, Harvard University Archives.

19. Ibid.

20. Hogan, *Of the Human Heart*, p. 70.

21. Cajori, "Teaching and History of Mathematics," pp. 140–141.

22. Steve Batterson, "Bôcher, Osgood, and the Ascendance of American Mathematics at Harvard," *Notices of the American Mathematical Society* 56 (September 2009), p. 918.

23. J. L. Coolidge, "William Elwood Byerly in Memoriam," *Bulletin of the American Mathematical Society* 42 (1936), pp. 295–298.

24. Charles W. Eliot, "Benjamin Peirce. I. Reminiscences of Peirce," *American Mathematical Monthly* 32 (January 1925), p. 3.

25. Daniel J. Cohen, *Equations from God* (Baltimore, Md.: Johns Hopkins University Press, 2007), pp. 66–67.

26. Julian Coolidge, "Mathematics 1870–1928," in *Development of Harvard University 1869–1929*, edited by Samuel Eliot Morison (Cambridge, Mass.: Harvard University Press, 1930), pp. 248–257.

27. W. E. Byerly, "Benjamin Peirce. III. Reminiscences," *American Mathematical Monthly* 32 (January 1925), p. 6.

28. Diann Renee Porter, "William Fogg Osgood at Harvard," doctoral thesis, University of Illinois at Chicago, 1997, p. 39.

29. Coolidge, "Mathematics 1870–1928," p. 253.

30. A. Lawrence Lowell, "Benjamin Peirce. II. Reminiscences," *American Mathematical Monthly* 32 (January 1925), p. 4.

31. Hogan, *Of the Human Heart*, p. 91.

32. Peterson, "Benjamin Peirce," pp. 94–95.

33. *Harvard Crimson*, April 26, 1883, available at http://www.thecrimson.com/article/1883/4/26/no-headline-a-story-is-told/.

34. J. L. Coolidge, "Three Hundred Years of Mathematics at Harvard," *American Mathematical Monthly* 50 (June–July 1943), p. 353.

35. Edward Hogan, "A Proper Spirit Is Abroad," *Historia Mathematica* 18 (1991), pp. 158–172.

36. Morison, *Three Centuries of Harvard*, pp. 305–307.

37. Cohen, *Equations from God*, pp. 64–65.

38. Ibid.

39. T. J. J. See, "The Services of Benjamin Peirce to American Mathematics and Astronomy," *Popular Astronomy* III (October 1895), pp. 50–51, HUG 300, Harvard University Archives.

40. Laura J. Snyder, *The Philosophical Breakfast Club* (New York: Random House Digital, 2011), p. 286.

41. Dirk Jan Struik, *Yankee Science in the Making* (Mineola, N.Y.: Dover, 1991), p. 416.

42. Hogan, *Of the Human Heart,* p. 27.
43. Batterson, "Bôcher, Osgood," p. 918.
44. Hogan, *Of the Human Heart,* pp. 25–26.
45. Ibid., p. 195.
46. James Clerk Maxwell, "Letter to William Thomson, August 1, 1857," in *The Scientific Letters and Papers of James Clerk Maxwell,* Vol. 1, *1846–1862,* edited by P. M. Harmon (Cambridge: Cambridge University Press, 1990), p. 527.
47. Matz, "Benjamin Peirce," p. 174.
48. Paul Weiss, "Peirce, Charles Sanders," in *Dictionary of American Biography* (1934), available at http://www.cspeirce.com/menu/library/aboutcsp/weissbio.htm.
49. Hogan, *Of the Human Heart,* p. 130.
50. Paul Meier and Sandy Zabell, "Benjamin Peirce and the Howland Will," *Journal of the American Statistical Association* 75 (September 1980), pp. 497–506.
51. Hogan, *Of the Human Heart,* pp. 286–287.
52. Hogan, "A Proper Spirit Is Abroad," p. 169.
53. I. Bernard Cohen (editor), *Benjamin Peirce: Father of Pure Mathematics in America* (New York: Arno Press, 1980), p. 280.
54. Ivor Grattan-Guinness and Alison Walsh, "Benjamin Peirce," in *Stanford Encyclopedia of Philosophy* (2008), available at http://plato.stanford.edu/entries/peirce-benjamin.
55. H. A. Newton, "Benjamin Peirce," *American Journal of Science,* 3rd series, **22** (September 1881), pp. 167–178, HUG 1680.154, Harvard University Archives.
56. Hogan, *Of the Human Heart,* p. 289.
57. Helena M. Pycior, "Benjamin Peirce's *Linear Associative Algebra,*" *Isis* 70 (1979), pp. 537–551.
58. Ibid.
59. Raymond C. Archibald, "Benjamin Peirce's Linear Associative Algebra and C. S. Peirce," *American Mathematical Monthly* **34** (December 1927), p. 526.
60. Pycior, "Peirce's *Linear Associative Algebra,*" p. 543.
61. Benjamin Peirce, "Linear Associative Algebra," *American Journal of Mathematics* **4** (1881), p. 109.
62. Ibid., p. 118.
63. Martin Czigler, "George David Birkhoff," in *Thinkers of the Twentieth Century,* edited by Roland Turner (Chicago: St. James Press, 1987), p. 80.
64. George D. Birkhoff, "Fifty Years of American Mathematics," in *Semicentennial Addresses of the American Mathematical Society,* Vol. 2 (New York: Arno Press, 1980).
65. Pycior, "Peirce's *Linear Associative Algebra,*" p. 551.

66. Hogan, *Of the Human Heart,* p. 298.

67. Grattan-Guinness and Walsh, "Benjamin Peirce."

68. Peirce, "Linear Associative Algebra," p. 97.

69. H. E. Hawkes, "Estimate of Peirce's Linear Associative Algebra," *American Journal of Mathematics* **24** (January 1902), pp. 87–95; Herbert Edwin Hawkes, "On Hypercomplex Number Systems," *Transactions of the American Mathematical Society* **3** (July 1902), pp. 312–330.

70. Newton, "Benjamin Peirce."

71. Peirce, "Linear Associative Algebra," p. 97.

72. Benjamin Peirce, "Address of Professor Benjamin Peirce, President of American Association for the Year 1853, on Retiring from the Duties of President." Available at http://projecteuclid.org/euclid.chmm/1263240516.

73. Cohen, *Equations from God*, p. 58.

74. Grattan-Guinness and Walsh, "Benjamin Peirce."

75. Benjamin Peirce, *Ideality in the Physical Sciences* (Boston: Little, Brown, 1881).

76. Pycior, "Peirce's *Linear Associative Algebra,*" p. 550.

77. Hogan, *Of the Human Heart,* p. 63.

78. Matz, "Benjamin Peirce," p. 178.

79. Hogan, *Of the Human Heart,* p. 9.

80. Moses King (editor), *Benjamin Peirce: A Memorial Collection* (Boston: Rand, Avery, 1881), p. 27, HUG 1680 145A, Box HD, Harvard University Archives.

81. Oliver Wendell Holmes, "Benjamin Peirce: Astronomer, Mathematician," *Atlantic Monthly* **46** (1880), p. 823, HUG 300, Harvard University Archives.

82. *Harvard Crimson,* October 15, 1880.

83. Coolidge, "Three Hundred Years of Mathematics at Harvard," p. 353.

2. Osgood, Bôcher, and the Great Awakening in American Mathematics

1. Steve Batterson, "Bôcher, Osgood, and the Ascendance of American Mathematics at Harvard," *Notices of the American Mathematical Society* **56** (September 2009), p. 916.

2. Ibid., p. 918.

3. Judith V. Grabiner, "Mathematics in America: The First Hundred Years," in *The Bicentennial Tribute to American Mathematics,* edited by Dalton Tarwater (Washington, D.C.: Mathematical Association of America, 1977), p. 19.

4. Karen Hunger Parshall, "Perspectives on American Mathematics," *Bulletin of the American Mathematical Society* **37** (2000), pp. 381–382.

5. Ibid.

6. Samuel Eliot Morison, *Three Centuries of Harvard* (Cambridge, Mass.: Harvard University Press, 1965), p. 356.

7. Karen Hunger Parshall and David E. Rowe, "American Mathematics Comes of Age: 1875–1900," in *A Century of Mathematics in America*, Part 3, edited by Peter Duren (Providence, R.I.: American Mathematical Society, 1989), pp. 7–8.

8. Grabiner, "Mathematics in America," p. 21.

9. Batterson, "Bôcher, Osgood," p. 919.

10. Parshall and Rowe, "American Mathematics Comes of Age," pp. 10–11.

11. Karen Hunger Parshall, "Eliakim Hastings Moore and the Founding of a Mathematical Community in America, 1892–1902," in *A Century of Mathematics in America*, Part 2, edited by Peter Duren (Providence, R.I.: American Mathematical Society, 1989), p. 156.

12. J. L. Coolidge, "Three Hundred Years of Mathematics at Harvard," *American Mathematical Monthly* 50 (June–July 1943), p. 353.

13. Julian L. Coolidge, George D. Birkhoff, and Edwin C. Kemble, "William Fogg Osgood," *Science* 98 (November 5, 1943), pp. 399–400.

14. U. G. Mitchell (editor), "Undergraduate Mathematical Clubs," *American Mathematical Monthly* 26 (June 1919), p. 263.

15. Batterson, "Bôcher, Osgood," pp. 922–923.

16. Joseph L. Walsh, "William Fogg Osgood," *Biographical Memoirs* 81 (2002), pp. 246–257.

17. Coolidge, Birkhoff, and Kemble, "William Fogg Osgood," pp. 399–400.

18. J. L. Walsh, "History of the Riemann Mapping Theorem," *American Mathematical Monthly* 80 (March 1973), pp. 270–276.

19. Bernard Osgood Koopman, "William Fogg Osgood—in Memoriam," *Bulletin of the American Mathematical Society* 50 (1944), pp. 139–142.

20. Garrett Birkhoff, "Mathematics at Harvard, 1836–1944," in *A Century of Mathematics in America*, Part 2, edited by Peter Duren (Providence, R.I.: American Mathematical Society, 1989), pp. 15–16.

21. J. L. Walsh, "William Fogg Osgood," in *A Century of Mathematics in America*, Part 2, edited by Peter Duren (Providence, R.I.: American Mathematical Society, 1989), p. 82.

22. Norbert Wiener, *Ex-Prodigy* (New York: Simon and Schuster, 1953), p. 232.

23. Diann Renee Porter, *William Fogg Osgood at Harvard*, doctoral thesis, University of Illinois at Chicago, 1997, p. 57.

24. J. L. Walsh, "William Fogg Osgood," in *Dictionary of American Biography*, Supplement 3, edited by John A. Garraty (New York: Scribner, 1973), pp. 575–576.

25. Walsh, "William Fogg Osgood," *Biographical Memoirs*.

26. William F. Osgood, "A Jordan of Positive Area," *Transactions of the American Mathematical Society* **4** (January 1903), pp. 107–112.

27. Raymond Clare Archibald, *A Semicentennial History of the American Mathematical Society, 1888–1938* (New York: American Mathematical Society, 1938), p. 155.

28. George D. Birkhoff, "The Progress of Science," *Scientific Monthly* **57** (November 1943), pp. 466–469.

29. G. D. Birkhoff, "Fifty Years of American Mathematics," in *Semicentennial Addresses of the American Mathematical Society*, Vol. 2 (New York: Arno Press, 1980), pp. 270–315.

30. Koopman, "William Fogg Osgood," pp. 139–142.

31. George D. Birkhoff, "The Scientific Work of Maxime Bôcher," *Bulletin of the American Mathematical Society* **25** (1919), pp. 197–215.

32. Birkhoff, "Fifty Years of American Mathematics."

33. Ibid.

34. J. D. Zund, "Maxime Bôcher," *American National Biography*, Vol. 3 (New York: Oxford University Press, 1999), pp. 88–89.

35. William F. Osgood, "Maxime Bôcher," *Bulletin of the American Mathematical Society* **35** (March–April 1929), pp. 205–217.

36. Archibald, *Semicentennial History*, pp. 162–163.

37. Birkhoff, "Scientific Work of Maxime Bôcher."

38. Maxime Bôcher, "The Fundamental Conceptions and Methods of Mathematics," address delivered before the department of mathematics of the International Congress of Arts and Science, St. Louis, September 20, 1904, *Bulletin of the American Mathematical Society* **11** (1904), pp. 115–135.

39. Ibid.

40. Ibid.

41. Garrett Birkhoff, "Some Leaders in American Mathematics: 1891–1941," in *The Bicentennial Tribute to American Mathematics*, edited by Dalton Tarwater (Washington, D.C.: Mathematical Association of America, 1977), pp. 33–34.

42. Walsh, "William Fogg Osgood," *Dictionary of American Biography*.

43. Archibald, *Semicentennial History*, p. 163.

44. "Maxime Bôcher," *Science* **48** (November 29, 1918), pp. 534–535.

45. J. Laurie Snell, "A Conversation with Joe Doob," *Statistical Science* **12** (November 1997), pp. 301–311.

46. Wiener, *Ex-Prodigy*, pp. 231–232.

47. Ibid.

48. Coolidge, Birkhoff, and Kemble, "William Fogg Osgood," pp. 399–400.

49. Julian Lowell Coolidge, "XV. Mathematics, 1870–1928," in *The Development of Harvard University*, edited by Samuel Eliot Morison (Cambridge, Mass.: Harvard University Press, 1930), p. 251.

50. Osgood, "Maxime Bôcher."
51. Angus Taylor, "A Life in Mathematics Remembered," *American Mathematical Monthly* **91** (December 1984), p. 607.
52. Birkhoff, "The Progress of Science," pp. 466–469.
53. D. V. Widder, "Some Mathematical Reminiscences," in *A Century of Mathematics in America*, Part 1, edited by Peter Duren (Providence, R.I.: American Mathematical Society, 1988), p. 80.
54. Walsh, "William Fogg Osgood," *Biographical Memoirs*.
55. "Maxime Bôcher," *Science* **18** (November 29, 1918), pp. 534–535.
56. "In the Death of Maxime Bôcher," *Transactions of the American Mathematical Society* **20** (January 1919).
57. Koopman, "William Fogg Osgood," pp. 139–142.
58. Coolidge, Birkhoff, and Kemble, "William Fogg Osgood," pp. 399–400.
59. Walsh, "William Fogg Osgood," *Biographical Memoirs*.
60. "Maxime Bôcher," pp. 534–535.
61. Archibald, *Semicentennial History*, p. 163.
62. William F. Osgood, "The Life and Services of Maxime Bôcher," *Bulletin of the American Mathematical Society* **25** (1919), pp. 337–350.
63. Records of the President of Harvard University, Abbott Lawrence Lowell, 1909–1933, General Correspondence, Series 1930–1933, letter, Lee to Lowell (August 26, 1932), folder 804, U115.160, Harvard University Archives.
64. Records of the President of Harvard University, Abbott Lawrence Lowell, 1909–1933, General Correspondence, Series 1930–1933, letter, Lowell to Osgood (October 13, 1932), folder 804, U115.160, Harvard University Archives.
65. David E. Zitarelli, "Towering Figures in Mathematics," *American Mathematical Monthly* **108** (August–September 2001), p. 617.
66. Batterson, "Bôcher, Osgood," p. 916.
67. Parshall, "Eliakim Hastings Moore," p. 155.

3. The Dynamical Presence of George David Birkhoff

1. Donald J. Albers and G. L. Alexanderson (editors), *Mathematical People: Profiles and Interviews* (Boston: Birkhäuser, 1985), p. 10.
2. H. S. Vandiver, "Some of My Recollections of George David Birkhoff," *Journal of Mathematical Analysis and Applications* **7** (1963), p. 272.
3. Albers and Alexanderson, *Mathematical People*, p. 10.
4. Garrett Birkhoff, "Mathematics at Harvard, 1836–1944," in *A Century of Mathematics in America*, Part 2, edited by Peter Duren (Providence, R.I.: American Mathematical Society, 1989), p. 25.
5. David E. Zitarelli, "Towering Figures in Mathematics," *American Mathematical Monthly* **108** (August–September 2001), p. 616.
6. Albers and Alexanderson, *Mathematical People*, p. 10.

7. Marston Morse, "George David Birkhoff and His Mathematical Work," *Bulletin of the American Mathematical Society* 52 (1946), pp. 357–391.

8. Garrett Birkhoff, "Some Leaders in American Mathematics: 1891–1941," in *The Bicentennial Tribute to American Mathematics: 1776–1976*, edited by Dalton Tarwater (Washington, D.C.: Mathematical Association of America, 1977), p. 40.

9. Morse, "George David Birkhoff."

10. Martin Czigler, "Birkhoff, George David," in *Thinkers of the Twentieth Century*, edited by Roland Turner (London: St. James Press, 1987), pp. 79–80.

11. June Barrow-Green, *Poincaré and the Three Body Problem* (Providence, R.I.: American Mathematical Society, 1997), p. 7.

12. Ibid., p. 15.

13. Henri Poincaré, *New Methods of Celestial Mechanics*, edited by Daniel L. Goroff (Woodbury, N.Y.: American Institute of Physics, 1993), p. xxi.

14. Barrow-Green, *Poincaré and the Three Body Problem*, pp. 22, 28.

15. Ibid., p. 1.

16. Poincaré, *New Methods of Celestial Mechanics*, p. 186.

17. Carol Parikh, *The Unreal Life of Oscar Zariski* (Boston: Academic Press, 1991), p. 40.

18. George D. Birkhoff, "Proof of Poincaré's Geometric Theorem," *Transactions of the American Mathematical Society* 14 (1913), pp. 14–22.

19. Poincaré, *New Methods of Celestial Mechanics*, p. 187.

20. Barrow-Green, *Poincaré and the Three Body Problem*, pp. 174, 223–225.

21. John Franks (Northwestern University), interview with the author, October 7, 2010.

22. Norbert Wiener, *Ex-Prodigy* (New York: Simon and Schuster, 1953), p. 230.

23. Birkhoff, "Mathematics at Harvard, 1836–1944," p. 27.

24. Morse, "George David Birkhoff."

25. Birkhoff, "Some Leaders in American Mathematics," p. 41.

26. George D. Birkhoff, "The Reducibility of Maps," *American Journal of Mathematics* 35 (April 1913), pp. 115–128.

27. J. J. O'Connor and E. F. Robertson, "History Topic: The Four Colour Theorem," *MacTutor History of Mathematics* (September 1996), available at http://www-groups.dcs.st-and.ac.uk/~history/PrintHT/The_four_colour_theorem.html.

28. G. L. Alexanderson, Review of "Four-Colors Suffice," *MAA Online Book Review Column* (December 12, 2003) available at http://mathdl.maa.org/mathDL/19/?pa=reviews&sa=viewBook&bookId=65760.

29. Vandiver, "Some of My Recollections," pp. 272–278.

30. Albers and Alexanderson, *Mathematical People*, pp. 12–13.

31. Rudolph Fritsch and Gerda Fritsch, *The Four-Color Theorem* (New York: Springer, 1998), p. 32.

32. George D. Birkhoff, "Dynamical Systems with Two Degrees of Freedom," *Proceedings of the National Academy of Sciences of the USA* **3** (April 1917), pp. 314–316.

33. Morse, "George David Birkhoff."

34. E. T. Whittaker, "Professor G. D. Birkhoff," *Nature* **154** (December 23, 1944), pp. 791–792.

35. Birkhoff, "Dynamical Systems." Marston Morse, "George David Birkhoff."

36. David Aubin, "George David Birkhoff, *Dynamical Systems* (1927), in *Landmark Writings in Western Mathematics, 1640–1940*, edited by L. Grattan-Guinness (Amsterdam: Elsevier, 2005), pp. 871–881.

37. Ibid.

38. Ibid.

39. G. D. Birkhoff, "What Is the Ergodic Theorem?," *American Mathematical Monthly* **49** (April 1942), pp. 222–226.

40. Franks, interview.

41. Joseph D. Zund, "George David Birkhoff and John von Neumann: A Question of Priority and the Ergodic Theorems, 1931–1932," *Historia Mathematica* **29** (2002), pp. 138–156.

42. Karen Hunger Parshall, "Perspectives on American Mathematics," *Bulletin of the American Mathematical Society* **37** (2000), p. 391.

43. Antti Knowles (Harvard University), interview with the author, October 15, 2010.

44. Ibid.

45. Franks, interview.

46. Terence Tao (UCLA), e-mail message to the author, October 16, 2010.

47. Ibid.

48. John von Neumann, "Letter to H. P. Robertson," in *John von Neumann: Selected Letters*, edited by Miklos Redei (Providence, R.I.: American Mathematical Society, 2005), pp. 208–210.

49. Ibid.

50. George D. Birkhoff, "Proof of the Ergodic Theorem," *Proceedings of the National Academy of Sciences of the USA* **17** (December 1931), pp. 656–660.

51. von Neumann, "Letter to H. P. Robertson."

52. G. D. Birkhoff and B. O. Koopman, "Recent Contributions to the Ergodic Theory," *Proceedings of the National Academy of Sciences of the USA* **18** (March 1932), pp. 279–282.

53. Zund, "George David Birkhoff and John von Neumann."

54. Steve Batterson, *Pursuit of Genius* (Wellesley, Mass.: A. K. Peters), p. 10.

55. Marston Morse, "George David Birkhoff."

56. Albers and Alexanderson, *Mathematical People*, p. 12.

57. Marjorie Van de Water, "Mathematical Measure for Art," *Science News Letter* **25** (March 17, 1934), pp. 170–172.

58. Ivars Peterson, "A Measure of Beauty," *Ivars Peterson's MathTrek* (May 24, 2004), available at http://www.maa.org/mathland/mathtrek_05_24_04.html.

59. J. Laurie Snell, "A Conversation with Joe Doob," available at http://www.dartmouth.edu/~chance/Doob/conversation.html.

60. Albers and Alexanderson, *Mathematical People,* p. 14.

61. Marston Morse, "George David Birkhoff."

62. Vandiver, "Some of My Recollections," pp. 273–274.

63. Edwin B. Wilson, "Obituary: George David Birkhoff," *Science* **102** (December 7, 1945), pp. 578–580.

64. R. E. Langer, "George David Birkhoff, 1884–1944," *Transactions of the American Mathematical Society* **60** (July 1946), pp. 1–2.

65. Marston Morse, "George David Birkhoff."

66. Oswald Veblen, "George David Birkhoff," in *Biographic Memoirs,* Vol. 80 (Washington, D.C.: The National Academy Press, 2001), pp. 1–14.

67. George D. Birkhoff, "Fifty Years of American Mathematics," in *Semicentennial Addresses of the American Mathematical Society,* Vol. 2 (New York: American Mathematical Society, 1938), pp. 270–315.

68. Constance Reid, *Courant* (New York: Springer, 1996), pp. 212–213.

69. Saunders Mac Lane, "Jobs in the 1930s and the Views of George D. Birkhoff," *Mathematical Intelligencer* **16** (1944), pp. 9–10.

70. Ioan James, *Remarkable Mathematicians: From Euler to von Neumann* (Cambridge: Cambridge University Press, 2002), p. 342.

71. Lipman Bers, "The Migration of European Mathematicians to America," in *A Century of Mathematics in America,* Part 1, edited by Peter Duran (Providence, R.I.: American Mathematical Society, 1988), p. 235.

72. Peter Lax (New York University), interview with the author, November 3, 2010.

73. Norbert Wiener, *I Am a Mathematician* (New York: Doubleday, 1956), pp. 28–31.

74. George Daniel Mostow (Yale University), interview with the author, April 9, 2010.

75. Wiener, *I Am a Mathematician,* pp. 28–31.

76. Nathan Reingold, "Refugee Mathematicians, 1933–1941," in *A Century of Mathematics in America,* Part 1, edited by Peter Duran (Providence, R.I.: American Mathematical Society, 1988), p. 183.

77. J. J. O'Connor and E. F. Robertson, "George David Birkhoff," *MacTutor History of Mathematics,* available at http://www-history.mcs.st-andrews.ac.uk/Biographies/Birkhoff.html.

78. Mac Lane. "Jobs in the 1930s."
79. Stephen H. Norwood, *The Third Reich in the Ivory Tower* (New York: Cambridge University Press, 2009), pp. 60–67.
80. Reid, *Courant,* p. 213.
81. Mostow, interview with the author, April 9, 2010.
82. Veblen, "George David Birkhoff."
83. Garrett Birkhoff, "The Rise of Modern Algebra, 1936 to 1960," in *Selected Papers on Algebra and Topology by Garrett Birkhoff,* edited by Gian-Carlo Rota and Joseph S. Oliveira (Boston: Birkhäuser, 1987), p. 586.
84. Bers, "Migration of European Mathematicians to America," pp. 233–234.
85. Reuben Hersh, "Under-represented Then Over-represented: A Memoir of Jews in American Mathematics," *College Mathematics Journal* **41** (January 2010), pp. 2–9.
86. Bers, "Migration of European Mathematicians to America," p. 242.

4. Analysis and Algebra Meet Topology

1. David E. Zitarelli, "Towering Figures in Mathematics," *American Mathematical Monthly* **108** (August–September 2001), pp. 606–635.
2. Calvin Moore, *Mathematics at Berkeley: A History* (Wellesley, Mass.: A. K. Peters, 2007), p. 63.
3. Joseph D. Zund, "Joseph Leonard Walsh," *American National Biography,* Vol. 22 (New York: Oxford University Press, 1999), pp. 571–572.
4. Joanne E. Snow and Colleen M. Hoover, "Mathematician as Artist: Marston Morse," *Mathematical Intelligencer* **32** (2010), pp. 11–18.
5. Marston Morse, "Relations between the Critical Points of a Real Function of n Independent Variables," *Transactions of the American Mathematical Society* **27** (1925), pp. 345–396.
6. Marston Morse, *The Calculus of Variations in the Large* (Providence, R.I.: American Mathematical Society, 1934).
7. S. Smale, "Marston Morse (1892–1977)," *Mathematical Intelligencer* **1** (March 1878), pp. 33–34.
8. Raoul Bott, "Marston Morse and His Mathematical Works," *Bulletin of the American Mathematical Society* **3** (November 1980), pp. 907–950.
9. Joseph D. Zund, "Marston Morse," *American National Biography,* Vol. 15 (New York: Oxford University Press, 1999), pp. 936–937.
10. Marston Morse, "Topology and Equilibria," edited by Abe Shenitzer and John Stillwell, *American Mathematical Monthly* **114** (November 2007), pp. 819–834.
11. Ibid.
12. Antti Knowles (Harvard University), interview with the author, November 30, 2010.

13. Morse, "Relations between the Critical Points."
14. Morse, "Topology and Equilibria."
15. Barry Mazur (Harvard University), interview with the author, November 24, 2010.
16. Snow and Hoover, "Mathematician as Artist."
17. Raoul Bott, "Marston Morse."
18. Ibid.
19. Saunders Mac Lane, *Saunders Mac Lane: A Mathematical Autobiography* (Wellesley, Mass.: A. K. Peters, 2005), p. 69.
20. Everett Pitcher, "Marston Morse," *Biographical Memoirs* 65 (Washington, D.C.: National Academy of Sciences, 1994), pp. 223–238.
21. Bott, "Marston Morse."
22. Pitcher, "Marston Morse."
23. Ibid.
24. Bott, "Marston Morse."
25. Snow and Hoover, "Mathematician as Artist."
26. Smale, "Marston Morse."
27. Stewart Cairns, "Marston Morse, 1892–1977," *Bulletin of the Institute of Mathematics Academia Sinica* 6 (October 1978), pp. i–ix.
28. Smale, "Marston Morse."
29. Bott, "Marston Morse."
30. Marston Morse, "Twentieth Century Mathematics," *American Scholar* 9 (1940), pp. 499–504.
31. Snow and Hoover, "Mathematician as Artist."
32. Marston Morse, "Mathematics and the Arts," in *Musings of the Masters*, edited by Raymond G. Ayoub (Washington, D.C.: Mathematical Association of America, 2004), p. 91.
33. Zund, "Marston Morse."
34. Garrett Birkhoff, "Some Leaders in American Mathematics: 1891–1941," in *The Bicentennial Tribute to American Mathematics: 1776–1976*, edited by Dalton Tarwater (Washington, D.C.: Mathematical Association of America, 1977), p. 69.
35. J. J. O'Connor and E. F. Robertson, "Hassler Whitney," *MacTutor History of Mathematics,* available at http://www-history.mcs.st-andrews.ac.uk/Biographies/Whitney.html.
36. "1985 Steele Prizes Awarded at Summer Meeting in Laramie," *Notices of the American Mathematical Society* 32 (1985), pp. 577–578.
37. Hassler Whitney, "Moscow 1935: Topology Moving Toward America," in *A Century of Mathematics in America,* Part 1, edited by Peter Duren (Providence, R.I.: American Mathematical Society, 1988), p. 99.
38. "1985 Steele Prizes Awarded."
39. Ron Bartlett, "Hassler Whitney: Humble about His World-wide Honor," *Princeton Packet,* March 2, 1983.

40. William Aspray and Albert Tucker, "Hassler Whitney (with Albert Tucker)," *Princeton Mathematics Community in the 1930s,* Transcript 43 (PMC43), 1985, available at http://www.princeton.edu/~mudd/finding_aids/mathoral/pmc43.htm.

41. Ibid.

42. Whitney, "Moscow 1935," p. 99.

43. Aspray and Tucker, "Hassler Whitney."

44. Shiing-shen Chern, "Hassler Whitney," *Proceedings of the American Philosophical Society* **138** (September 1994), pp. 465–467.

45. Karen Hunger Parshall, "Perspectives on American Mathematics," *Bulletin of the American Mathematical Society* **37** (2000), p. 391.

46. Chern, "Hassler Whitney."

47. Ibid.

48. Jean Dieudonné, *A History of Algebraic and Differential Topology* (Boston: Birkhäuser, 1989), pp. 78–79.

49. John W. Milnor and James D. Stasheff, *Characteristic Classes* (Princeton, N.J.: Princeton University Press, 1974), p. v.

50. Hassler Whitney, *Geometric Integration Theory* (Princeton, N.J.: Princeton University Press, 1957).

51. Whitney, "Moscow 1935," p. 117.

52. Ibid.

53. Alex Heller, "Samuel Eilenberg," *Biographical Memoirs,* Vol. 79 (Washington, D.C.: National Academy Press, 2001), pp. 119–122.

54. Mac Lane, *Mathematical Autobiography,* pp. 22, 31.

55. Ibid., p. 33.

56. Ibid., p. 53.

57. Philip J. Davis, "Mister Mathematics: Saunders Mac Lane," *SIAM News,* November 20, 2005.

58. Saunders Mac Lane, "Mathematics at Göttingen under the Nazis," *Notices of the American Mathematical Society* **42** (October 1995), pp. 1134–1138.

59. Ivan Niven, "The Threadbare Thirties," in *A Century of Mathematics in America,* Part 1, edited by Peter Duren (Providence, R.I.: American Mathematical Society, 1988), p. 219.

60. G. L. Alexanderson, "A Conversation with Saunders Mac Lane," *College Mathematics Journal* **20** (January 1989), pp. 2–25.

61. Mac Lane, *Mathematical Autobiography,* p. 135.

62. Ibid., p. 72.

63. Gaston de los Reyes, "Fellows Promote Genius, Continue Tradition," *Harvard Crimson,* March 5, 1994.

64. Saunders Mac Lane, "Garrett Birkhoff and the 'Survey of Modern Algebra,'" *Notices of the American Mathematical Society* **44** (December 1997), pp. 1438–1439.

65. Mac Lane, *Mathematical Autobiography*, p. 137.
66. Alexanderson, "Conversation with Saunders Mac Lane."
67. Barry Mazur (Harvard University), interview with the author, December 9, 2009.
68. Mac Lane, *Mathematical Autobiography*, p. xi.
69. Ibid., p. 101.
70. S. Mac Lane, "Samuel Eilenberg and Categories," *Journal of Pure and Applied Algebra* **168** (2002), pp. 127–131.
71. Hyman Bass, Henri Cartan, Peter Freyd, Alex Heller, and Saunders Mac Lane, "Samuel Eilenberg (1913–1998)," *Notices of the American Mathematical Society* **45** (November 1998), pp. 1344–1352.
72. Alexanderson, "Conversation with Saunders Mac Lane."
73. Michael Barr (McGill University), interview with the author, November 29, 2010.
74. Mac Lane, *Mathematical Autobiography*, pp. 125–126.
75. Saunders Mac Lane, "Concepts and Categories in Perspective," in *A Century of Mathematics in America*, Part 1, edited by Peter Duren (Providence, R.I.: American Mathematical Society, 1988), pp. 323–365.
76. Ibid.
77. Alexandra C. Bell, "Ex-Math Prof Mac Lane, 95, Dies," *Harvard Crimson*, April 25, 2005.
78. Steve Koppes, "Saunders Mac Lane, Mathematician, 1909–2005," press release, University of Chicago News Office, April 21, 2005.
79. Steve Awodey, "In Memoriam: Saunders Mac Lane," *Bulletin of Symbolic Logic* **13** (March 2007), pp. 115–119.
80. Colin McLarty (Case Western Reserve University), interview with the author, November 30, 2010.
81. Colin McLarty (Case Western Reserve University), interview with the author, December 12, 2010.
82. William Lawvere (SUNY Buffalo), interview with the author, November 29, 2010.
83. Davis, "Mister Mathematics."
84. Della Fenster, "Reviews," *American Mathematical Monthly* **113** (December 2006), pp. 947–951.
85. Davis, "Mister Mathematics."
86. Walter Tholen, "Saunders Mac Lane, 1909–2005: Meeting a Grand Leader," *Scientiae Mathematicae Japonicae* **63** (January 2006), pp. 13–24.

5. Analysis Most Complex

1. Albert Marden (University of Minnesota), e-mail message to the author, August 6, 2011.

2. Olli Lehto, "On the Life and Work of Lars Ahlfors," *Mathematical Intelligencer* 20 (1985), pp. 4–8.

3. Lars Ahlfors, "Author's Preface," in *Lars Valerian Ahlfors, Collected Papers,* Vol. 1, *1929–1955,* edited by Rae Michael Shortt (Boston: Birkhäuser, 1982), p. xi.

4. Donald J. Albers, "An Interview with Lars V. Ahlfors," *College Mathematics Journal* 29 (March 1998), pp. 82–92.

5. Ahlfors, "Author's Preface," p. xi.

6. Albers, "Interview with Lars V. Ahlfors."

7. Troels Jorgensen, in "Lars Valerian Ahlfors," edited by Steven G. Krantz, *Notices of the American Mathematical Society* 45 (February 1998), pp. 248–255.

8. Albers, "Interview with Lars V. Ahlfors."

9. Lars Ahlfors, "The Joy of Function Theory," in *A Century of Mathematics in America,* Part 3, edited by Peter Duren (Providence, R.I.: American Mathematical Society, 1989), pp. 443–447.

10. Ahlfors, "Joy of Function Theory."

11. Osmo Pekonen, "Lars Ahlfors, Finland's Greatest Mathematician," *CSCnews* (*Information Technology for Science in Finland*) 2 (2007), p. 39.

12. Albers, "Interview with Lars V. Ahlfors."

13. Ahlfors, "Joy of Function Theory."

14. Aimo Hinkkanen (University of Illinois), interview with the author, June 13, 2011.

15. Ahlfors, "Author's Preface," p. xii.

16. Albers, "Interview with Lars V. Ahlfors."

17. Ibid.

18. Ahlfors, "Joy of Function Theory."

19. Ahlfors, "Author's Preface," p. xii.

20. Lehto, "Life and Work of Lars Ahlfors."

21. Robert Osserman, "The Geometry Renaissance in America: 1938–1988," in *A Century of Mathematics in America,* Part 1, edited by Peter Duren (Providence, R.I.: American Mathematical Society, 1988), p. 513.

22. Albers, "Interview with Lars V. Ahlfors."

23. Lehto, "Life and Work of Lars Ahlfors."

24. David Drasin (Purdue University), interview with the author, June 15, 2011.

25. David Drasin, e-mail message to the author, June 15, 2011.

26. Ibid.

27. Hinkkanen, interview.

28. L. V. Ahlfors, "Zur Theorie der Überlagerungsflächen," *Acta Mathematica* 65 (1935), pp. 157–194.

29. Hinkkanen, interview.

30. David Drasin, e-mail message to the author, June 14, 2011.
31. Robert Osserman, "Conformal Geometry," in "The Mathematics of Lars Valerian Ahlfors," edited by Steven G. Krantz, *Notices of the American Mathematical Society* 45 (February 1998), pp. 233–236.
32. Albers, "Interview with Lars V. Ahlfors."
33. Osserman, "Conformal Geometry."
34. Ahlfors, "Author's Preface," p. xii.
35. Albers, "Interview with Lars V. Ahlfors."
36. L. V. Ahlfors, "The Theory of Meromorphic Curves," *Acta Societas Scientiarum Fennicae* 3 (1941), pp. 3–31.
37. Frederick Gehring, "Lars Valerian Ahlfors," *Biographical Memoirs* 87 (2005), pp. 1–27.
38. Hung-Hsi Wu (University of California, Berkeley), e-mail message to the author, July 18, 2011.
39. Lehto, "Life and Work of Lars Ahlfors."
40. Albers, "Interview with Lars V. Ahlfors."
41. Ibid.
42. Steven G. Krantz, *Mathematical Apocrypha* (Washington, D.C.: Mathematical Association of America, 2002), p. 166.
43. Pekonen, "Lars Ahlfors," p. 39.
44. Ahlfors, "Author's Preface," p. xiv.
45. Ibid.
46. Raoul Bott, in "Lars Valerian Ahlfors," edited by Steven G. Krantz, *Notices of the American Mathematical Society* 45 (February 1998), pp. 254–255.
47. Marden, e-mail message.
48. L. Ahlfors and A. Beurling, "Conformal Invariants and Function-Theoretic Null Sets," *Acta Mathematica* 83 (1950), pp. 101–129.
49. Steven G. Krantz (editor), "Lars Valerian Ahlfors," *Notices of the American Mathematical Society* 45 (February 1998), pp. 248–249.
50. Marden, e-mail message.
51. Hinkkanen, interview.
52. Frederick Gehring, "Quasiconformal Mappings," in "The Mathematics of Lars Valerian Ahlfors," edited by Steven G. Krantz, *Notices of the American Mathematical Society* 45 (February 1998), pp. 239–242.
53. Lehto, "Life and Work of Lars Ahlfors."
54. Clifford Earle (Cornell University), interview with the author, June 10, 2011.
55. Marden, e-mail message.
56. Lehto, "Life and Work of Lars Ahlfors."
57. Gehring, "Quasiconformal Mappings."
58. A. Beurling and L. Ahlfors, "The Boundary Correspondence under Quasiconformal Mappings," *Acta Mathematica* 96 (1956), pp. 125–142.

59. Lars Ahlfors, "The Story of a Friendship: Recollections of Arne Beurling," *Mathematical Intelligencer* **15** (1993), pp. 25–27.

60. Gehring, "Quasiconformal Mappings."

61. David Drasin, e-mail message to the author, June 20, 2011.

62. Dennis Hejhal (University of Minnesota), interview with the author, July 5, 2011.

63. Marden, e-mail message.

64. Andrew Gleason, George Mackey, and Raoul Bott, "Faculty of Arts and Sciences—Memorial Minute: Lars Valerian Ahlfors," *Harvard Gazette,* January 24, 2001.

65. J. J. O'Connor and E. F. Robertson, "Lipman Bers," *MacTutor History of Mathematics,* available at http://www-history.mcs.st-andrews.ac.uk/Biographies/Bers.html.

66. William Abikoff, "Lipman Bers," *Notices of the American Mathematical Society* **72** (1995), pp. 385–404.

67. Lars Ahlfors and Lipman Bers, "Riemann's Mapping Theorem for Variable Metrics," *Annals of Mathematics* **72** (September 1960), pp. 385–404.

68. Lars V. Ahlfors, "Quasiconformal Mappings, Teichmüller Spaces, and Kleinian Groups," *Proceedings of the International Congress of Mathematicians* (Helsinki, 1978), pp. 71–84, available at http://www.mathunion.org/ICM/ICM1978.1/Main/icm1978.1.0071.0084.ocr.pdf.

69. John Willard Milnor, *Dynamics in One Complex Variable,* 3rd ed. (Princeton, N.J.: Princeton University Press, 2006), p. 165.

70. Ahlfors, "Quasiconformal Mappings."

71. Irwin Kra (Stony Brook University), interview with the author, June 15, 2011.

72. Ibid.

73. Lars V. Ahlfors, "Finitely Generated Kleinian Groups," *American Journal of Mathematics* **86** (April 1964), pp. 413–429.

74. Irwin Kra, "Kleinian Groups," in "Lars Valerian Ahlfors," edited by Steven G. Krantz, *Notices of the American Mathematical Society* **45** (February 1998), pp. 248–255.

75. Kra, interview.

76. Irwin Kra, "Creating an American Mathematical Tradition: The Extended Ahlfors-Bers Family," in *A Century of Mathematical Meetings,* edited by Bettye Anne Case (Providence, R.I.: American Mathematical Society, 1996), pp. 265–280.

77. Marden, e-mail message.

78. Ahlfors, "Finitely Generated Kleinian Groups."

79. Kra, "Creating an American Mathematical Tradition."

80. Lehto, "Life and Work of Lars Ahlfors."

81. Krantz, "Lars Valerian Ahlfors," pp. 248–249.

82. Marden, e-mail message.
83. Albers, "Interview with Lars V. Ahlfors."
84. Bott, in "Lars Valerian Ahlfors," pp. 254–255.
85. Lehto, "Life and Work of Lars Ahlfors."
86. "The 1981 Wolf Foundation Prize in Mathematics," available at http://www.wolffund.org.il/index.php?dir=site&page=winners&cs=237&language=eng.
87. Hejhal, interview.
88. Krantz, "Lars Valerian Ahlfors," pp. 248–249.
89. Gleason, Mackey, and Bott, "Faculty of Arts and Sciences."
90. Hejhal, interview.
91. Dennis Hejhal, in "Lars Valerian Ahlfors," edited by Steven G. Krantz, *Notices of the American Mathematical Society* 45 (February 1998), pp. 251–253.
92. Robert Osserman, "Lars Valerian Ahlfors (1907–1996)," edited by Steven G. Krantz, *Notices of the American Mathematical Society* 45 (February 1998), p. 250.
93. Marden, e-mail message.
94. Lehto, "Life and Work of Lars Ahlfors."

6. The War and Its Aftermath

1. Saunders Mac Lane, *Saunders Mac Lane: A Mathematical Autobiography* (Wellesley, Mass.: A. K. Peters, 2005), p. 120.
2. John McCleary, "Airborne Weapons Accuracy," *Mathematical Intelligencer* 28 (2006), pp. 17–21.
3. Garrett Birkhoff, "Mathematics at Harvard, 1836–1944," in *A Century of Mathematics in America,* Part 2, edited by Peter Duren (Providence, R.I.: American Mathematical Society, 1989), pp. 3–58.
4. Ibid.
5. Garrett Birkhoff, "The Rise of Modern Algebra, 1936 to 1950," in *Selected Papers on Algebra and Topology by Garrett Birkhoff,* edited by Gian-Carlo Rota and Joseph S. Oliveira (Boston: Birkhäuser, 1987), pp. 585–605.
6. Necia Grant Cooper, Roger Eckhardt, and Nancy Shera, *From Cardinals to Chaos: Reflections on the Life and Legacy of Stanislaw Ulam* (New York: Cambridge University Press, 1989).
7. J. Barkley Rosser, "Mathematics and Mathematicians in World War II," in *A Century of Mathematics in America,* Part 1, edited by Peter Duren (Providence, R.I.: American Mathematical Society, 1988), pp. 303–309; John T. Bethell, *Harvard Observed* (Cambridge, Mass.: Harvard University Press, 1998), p. 147.
8. Rosser, "Mathematics and Mathematicians."
9. Ibid.

10. Ethan Bolker (editor), "Andrew M. Gleason, 1921–2008," *Notices of the American Mathematical Society* **56** (November 2009), pp. 1236–1239.

11. Andrew M. Gleason, "Andrew M. Gleason," in *More Mathematical People: Contemporary Conversations,* edited by Donald J. Albers, Gerald L. Alexanderson, and Constance Reid (San Diego: Academic Press, 1990), p. 87.

12. George Dyson, "Turing Centenary: The Dawn of Computing," *Nature* **482** (February 23, 2012), pp. 459–460.

13. Ethan Bolker (University of Massachusetts–Boston), interview with the author, February 1, 2012.

14. John Burroughs, David Lieberman, and Jim Reeds, "The Secret Life of Andy Gleason," *Notices of the American Mathematical Society* **56** (November 2009), pp. 1239–1243.

15. Deborah Hughes Hallet, "Andy Gleason: Teacher," *Notices of the American Mathematical Society* **56** (November 2009), p. 1264.

16. Burroughs et al., "Secret Life of Andy Gleason," p. 1240.

17. Richard Ruggles and Henry Brodie, "An Empirical Approach to Economic Intelligence in World War II," *Journal of the American Statistical Association* **42 B** (March 1947), pp. 72–91.

18. "Technology History," Bletchley Park Science and Innovation Center, available at http://www.bpsic.com/bletchley-park/technology-history.

19. Burroughs et al., "Secret Life of Andy Gleason."

20. Benedict Gross, David Mumford, and Barry Mazur, "Andrew Mattei Gleason," *Harvard Gazette,* April 1, 2010.

21. Gleason, "Andrew M. Gleason," p. 88.

22. John Wermer, "Gleason's Work on Banach Algebras," *Notices of the American Mathematical Society* **56** (November 2009), pp. 1248–1251.

23. Joel Spencer, "Andrew Gleason's Discrete Mathematics," *Notices of the American Mathematical Society* **56** (November 2009), pp. 1251–1253.

24. Gleason, "Andrew M. Gleason," p. 89.

25. Bolker, interview.

26. R. L. Graham, "Roots of Ramsey Theory," in *Andrew M. Gleason: Glimpses of a Life in Mathematics,* edited by Ethan Bolker (University of Massachusetts at Boston, 1992), pp. 39–47.

27. R. E. Greenwood and A. M. Gleason, "Combinatorial Relations and Chromatic Graphs," *Canadian Journal of Mathematics* **7** (1955), pp. 1–7.

28. Spencer, "Andrew Gleason's Discrete Mathematics."

29. B. D. McKay and S. P. Radziszowski, "R(4,5)=25," *Journal of Graph Theory* **19** (May 1995), pp. 309–322.

30. Joel Spencer, *Ten Lectures on the Probabilistic Method* (Philadelphia: Society for Industrial and Applied Mathematics, 1987), p. 4.

31. George W. Mackey, "Remarks at Andrew Gleason's Retirement Conference," in *Andrew M. Gleason: Glimpses of a Life in Mathematics,* edited by Ethan Bolker (University of Massachusetts at Boston, 1992), pp. 60–61.

32. Gleason, "Andrew M. Gleason," p. 91.

33. David Hilbert, "Mathematical Problems," *Bulletin of the American Mathematical Society* 8 (1902), pp. 437–479.

34. Benjamin H. Yandell, *The Honors Class: Hilbert's Problems and Their Solvers* (Natick, Mass.: A. K. Peters, 2002), p. 144.

35. Ibid.

36. Richard Palais, "Gleason's Contribution to the Solution of Hilbert's Fifth Problem," *Notices of the American Mathematical Society* 56 (November 2009), pp. 1243–1247.

37. Ibid.

38. Richard Palais (University of California, Irvine), interview with the author, September 5, 2011.

39. Richard Palais, e-mail message to the author, February 16, 2012.

40. Yandell, *The Honors Class,* p. 152.

41. G. Daniel Mostow, *Science at Yale: Mathematics* (New Haven, Conn.: Yale University, 2001), p. 22.

42. Yandell, *The Honors Class,* pp. 152–153.

43. Deane Montgomery and Leo Zippin, "Small Subgroups of Finite-Dimensional Groups," *Annals of Mathematics* 56 (September 1952), pp. 213–241.

44. Andrew M. Gleason, "Groups without Small Subgroups," *Annals of Mathematics* 58 (1953), pp. 193–212.

45. Richard Palais, interview with the author, September 5, 2011.

46. Mostow, *Science at Yale,* p. 22.

47. Palais, e-mail message.

48. Gleason, "Andrew M. Gleason," pp. 92–93.

49. Jeremy J. Gray, *The Hilbert Challenge* (New York: Oxford University Press, 2000), p. 178.

50. Palais, interview.

51. Paul Chernoff, "Andy Gleason and Quantum Mechanics," *Notices of the American Mathematical Society* 56 (November 2009), pp. 1253–1259.

52. Hallet et al., "Andy Gleason: Teacher."

53. Sheldon Gordon, "Second Star on the Right and Straight on to Morning," in *Andrew M. Gleason: Glimpses of a Life in Mathematics,* edited by Ethan Bolker (University of Massachusetts at Boston, 1992), p. 55.

54. Bolker, "Andrew M. Gleason."

55. Andrew M. Gleason, "Angle Trisection, the Heptagon, and the Triskaidecagon," *American Mathematical Monthly* 95 (March 1988), pp. 185–194.

56. Bolker, interview.

57. Bolker, "Andrew M. Gleason."
58. Roger Howe (Yale University), interview with the author, August 2, 2011.
59. V. S. Varadarajan, "George Mackey and His Work on Representation Theory and Foundations of Physics," *Contemporary Mathematics* **449** (2008), pp. 417–446.
60. Veeravalli S.Varadarajan (UCLA), interview with the author, January 12, 2012.
61. Chernoff, "Andy Gleason and Quantum Mechanics."
62. Mackey, "Andrew Gleason's Retirement Conference."
63. Chernoff, "Andy Gleason and Quantum Mechanics."
64. Andrew M. Gleason, "Measures on the Closed Subspaces of a Hilbert Space," *Journal of Mathematics and Mechanics* **6** (1957), pp. 885–893.
65. Varadarajan, interview.
66. Howe, interview.
67. Chernoff, "Andy Gleason and Quantum Mechanics," p. 1253.
68. Howe, interview.
69. Mackey, "Andrew Gleason's Retirement Conference."
70. Ibid.
71. Judith A. Packer, "George Mackey: A Personal Remembrance," *Notices of the American Mathematical Society* **54** (August 2007), pp. 837–841.
72. Caroline Series, "George Mackey," *Notices of the American Mathematical Society* **54** (August 2007), pp. 844–847.
73. Ann Mackey, "Eulogy for My Father, George Mackey," *Notices of the American Mathematical Society* **54** (August 2007), pp. 849–850.
74. Bryan Marquard, "George Mackey, Professor Devoted to Truth, Theorems," *Boston Globe*, April 28, 2006.
75. Packer, "George Mackey."
76. Series, "George Mackey."
77. Roger Howe (Yale University), interview with the author, September 6, 2011.
78. G. W. Mackey, Letter to Stephanie Singer, September 19, 1982, available at http://www.symmetrysinger.com/Mackey/letter1.pdf.
79. Varadarajan, interview.
80. Howe, interview.
81. David Mumford, "George Whitelaw Mackey," *Proceedings of the American Philosophical Society* **152** (December 2008), pp. 559–663.
82. Andrew Gleason, Calvin Moore, David Mumford, Clifford Taubes, and Shlomo Sternberg, "George Whitelaw Mackey," *Harvard Gazette*, December 18, 2008.
83. Mackey, "Eulogy for My Father."
84. Jean Berko Gleason, "A Life Well Lived," *Notices of the American Mathematical Society* **56** (November 2009), pp. 1266–1267.

85. Bolker, "Andrew M. Gleason," p. 1261.
86. Spencer, "Andrew Gleason's Discrete Mathematics," p. 1252.
87. Ibid.
88. Ethan Bolker, ". . . and the Work Is Play . . ," in *Andrew M. Gleason: Glimpses of a Life in Mathematics,* edited by Ethan Bolker (University of Massachusetts at Boston, 1992), pp. 39–47.

7. The Europeans

1. Nathan Reingold, "Refugee Mathematicians in the United States of America, 1933–1941: Reception and Reaction," in *A Century of Mathematics in America,* Vol. 1, edited by Peter Duren (Providence, R.I.: American Mathematical Society, 1988), pp. 175–200.
2. Carol Parikh, *The Unreal Life of Oscar Zariski* (Boston: Academic Press, 1991), p. 10.
3. H. Hironaka, G. Mackey, D. Mumford, and J. Tate, "Oscar Zariski: Memorial Minute Adopted by the Faculty of Arts and Sciences," *Harvard University Gazette* 83 (May 1988).
4. Parikh, *Unreal Life of Oscar Zariski,* p. 14.
5. Oscar Zariski, "Preface," in *Oscar Zariski: Collected Papers,* Vol. 4, edited by J. Lipman and B. Teissier (Cambridge, Mass.: MIT Press, 1979), pp. xi–xviii.
6. Parikh, *Unreal Life of Oscar Zariski,* p. 16.
7. Ibid., p. 20.
8. Ibid., p. 25.
9. Ibid., p. 25.
10. Zariski, "Preface."
11. Ibid.
12. Ibid.
13. "Mathematician, Oscar Zariski, Dead at 86," *Harvard Crimson,* July 11, 1986.
14. Ioan James, *Remarkable Mathematicians: From Euler to von Neumann* (Cambridge: Cambridge University Press, 2007), p. 403.
15. "Mathematician, Oscar Zariski."
16. Parikh, *Unreal Life of Oscar Zariski,* pp. 69–70.
17. Zariski, "Preface."
18. Parikh, *Unreal Life of Oscar Zariski,* p. 76.
19. Ibid., p. 79.
20. Ibid., p. 120.
21. Allyn Jackson, "Interview with Heisuke Hironaka," *Notices of the American Mathematical Society* 52 (October 2005), pp. 1010–1019.
22. Heisuke Hironaka (Harvard University), interview with the author, March 31, 2011.

23. Garrett Birkhoff, "Oscar Zariski (24 April 1899–4 July 1986)," *Proceedings of the American Philosophical Society* 137 (1993), pp. 307–320.

24. Parikh, *Unreal Life of Oscar Zariski*, p. 103.

25. D. Mumford, "Oscar Zariski, 1899–1986," *Notices of the American Mathematical Society* 33 (1986), pp. 891–894.

26. David Mumford (Harvard University), interview with the author, March 11, 2011.

27. Oscar Zariski, "Foundations of a General Theory of Birational Correspondences," *Transactions of the American Mathematical Society* 53 (1943), pp. 490–542.

28. Oscar Zariski, "Theory and Applications of Holomorphic Functions on Algebraic Varieties over Arbitrary Ground Fields," *Memoirs of the American Mathematical Society* 5 (1951), pp. 1–90; Mumford, "Oscar Zariski."

29. Michael Artin (MIT), interview with the author, March 31, 2011.

30. Mumford, interview.

31. Mumford, "Oscar Zariski."

32. Zariski, "Preface."

33. Parikh, *Unreal Life of Oscar Zariski*, p. 95.

34. Hironaka et al., "Oscar Zariski."

35. Leila Schneps, "Review: Grothendieck-Serre Correspondence," *Mathematical Intelligencer* 29 (2007), pp. 1–8.

36. Pierre Colmez and Jean-Pierre Serre, *Grothendieck-Serre Correspondence* (Providence, R.I.: American Mathematical Society, 2004), p. 114.

37. Steven L. Kleiman, "Steven L. Kleiman," in *Recountings: Conversations with MIT Mathematicians*, edited by Joel Segel (Wellesley, Mass.: A. K. Peters, 2009), pp. 278–279.

38. J. Lipman, "Oscar Zariski," *Progress in Mathematics* 181 (2000), pp. 1–4.

39. Parikh, *Unreal Life of Oscar Zariski*, p. 119.

40. Ibid.

41. Hironaka et al., "Oscar Zariski."

42. Jackson, "Interview with Heisuke Hironaka."

43. Ibid.

44. Mumford, "Oscar Zariski."

45. Ibid.

46. Parikh, *Unreal Life of Oscar Zariski*, p. 165.

47. Arthur P. Mattuck, "Arthur P. Mattuck," in *Recountings: Conversations with MIT Mathematicians*, edited by Joel Segel (Wellesley, Mass.: A. K. Peters, 2009), pp. 52–54.

48. Heisuke Hironaka, "Zariski's Papers on Resolution of Singularities," in *Oscar Zariski: Collected Papers*, edited by H. Hironaka and D. Mumford (Cambridge, Mass.: MIT Press, 1972), pp. 223–231.

49. Ibid.

50. S. S. Abhyankar, "Local Uniformization on Algebraic Surfaces over Ground Fields of $p \neq 0$," *Annals of Mathematics* 63 (May 1956), pp. 491–526.

51. Parikh, *Unreal Life of Oscar Zariski*, p. 163.

52. Jackson, "Interview with Heisuke Hironaka."

53. Hironaka, interview.

54. Ibid.

55. Mumford, interview.

56. David Mumford, "Autobiography of David Mumford," in *Fields Medallists' Lectures*, edited by Michael Atiyah and Daniel Iagolnitzer (Singapore: World Scientific, 2003), p. 233.

57. Mumford, interview.

58. David Mumford, "Autobiography of David Mumford," Shaw Prize 2006, available at http://www.shawprize.org/en/shaw.php?tmp=3&twoid=51&threeid=63&fourid=107&fiveid=24.

59. David Gieseker (UCLA), interview with the author, June 8, 2011.

60. David Mumford, e-mail communiqué to the author, March 11, 2011.

61. Mumford, interview.

62. Mumford, e-mail communiqué.

63. Michael Artin, "Michael Artin," in *Recountings: Conversations with MIT Mathematicians*, edited by Joel Segel (Wellesley, Mass.: A. K. Peters, 2009), pp. 358–359.

64. Michael Artin (MIT), interview with the author, March 7, 2011.

65. Ibid.

66. Parikh, *Unreal Life of Oscar Zariski*, p. 157.

67. Ibid.

68. "1981 Steele Prizes," *Notices of the American Mathematical Society* 28 (1981), pp. 504–507.

69. Wolf Foundation, "The 1981 Wolf Foundation Prize in Mathematics," available at http://www.wolffund.org.il/index.php?dir=site&page=winners&cs=241.

70. Lipman, "Oscar Zariski."

71. Parikh, *Unreal Life of Oscar Zariski*, p. 165.

72. Ibid., p. 171.

73. Richard Brauer, "Preface," in *Collected Papers*, Vol. 1, edited by Paul Fong and Warren J. Wong (Cambridge, Mass.: MIT Press, 1980), pp. xv–xix.

74. Ibid.

75. Ibid.

76. Walter Feit, "Richard D. Brauer," *Bulletin of the American Mathematical Society* 1 (January 1979), pp. 1–20.

77. Ibid.

78. J. A. Green, "Richard Dagobert Brauer," *Bulletin of the London Mathematical Society* 10 (1978), pp. 317–342.

79. Feit, "Richard D. Brauer."

80. Charles W. Curtis, *Pioneers of Representation Theory* (Providence, R.I.: American Mathematical Society, 1999), p. 205.

81. Jonathan L. Alperin, "Brauer, Richard Dagobert," *American National Biography Online*, Oxford University Press, available at http://www.anb.org.

82. Mario Livio, *The Equation That Couldn't Be Solved* (New York: Simon and Schuster, 2005), p. 260.

83. Charles Curtis (University of Oregon), interview with author. March 26, 2012.

84. Ibid.

85. Ibid.

86. Charles W. Curtis, "Richard Brauer: Sketches from His Life and Work," *American Mathematical Monthly* 110 (October 2003), pp. 665–678;

87. Richard Brauer and K. A. Fowler, "On Groups of Even Order," *Annals of Mathematics* 62 (November 1955), pp. 565–583; Green, "Richard Dagobert Brauer."

88. Curtis, *Pioneers of Representation Theory*, p. 208.

89. Feit, "Richard D. Brauer."

90. Walter Feit and John G. Thompson, "Solvability of Groups of Odd Order," *Pacific Journal of Mathematics* 13 (1963), pp. 775–1029.

91. John T. Tate, George W. Mackey, and Barry C. Mazur, "Richard Dagobert Brauer: Memorial Minute," *Harvard University Gazette* 75 (February 1, 1980).

92. Livio, *The Equation That Couldn't Be Solved*, p. 259.

93. Stephen Ornes, "Prize Awarded for Largest Mathematical Proof," *New Scientist*, September 9, 2011.

94. Daniel Gorenstein, "The Classification of Finite Simple Groups. I. Simple Groups and Local Analysis," *Bulletin of the American Mathematical Society* 1 (January 1979), pp. 43–199.

95. Feit, "Richard D. Brauer."

96. Green, "Richard Dagobert Brauer."

97. Allyn Jackson, "Interview with Raoul Bott," *Notices of the American Mathematical Society* 48 (April 2001), pp. 374–382.

98. Harold M. Edwards, in "Raoul Bott as We Knew Him," in *A Celebration of the Mathematical Legacy of Raoul Bott*, edited by P. Robert Kotiuga (Providence, R.I.: American Mathematical Society), pp. 43–50.

99. Raoul Bott, "Autobiographical Sketch," in *Raoul Bott Collected Works*, Vol. 1, edited by Robert D. MacPherson (Boston: Birkhäuser, 1994), pp. 3–10.

100. Loring Tu, "The Life and Works of Raoul Bott," *Notices of the American Mathematical Society* 53 (May 2006), pp. 554–570.

101. R. Bott and R. J. Duffin, "Impedance Synthesis without Use of Transformers," *Journal of Applied Physics* 20 (1949), p. 816.

102. Raoul Bott, "Marston Morse and His Mathematical Works," *Bulletin of the American Mathematical Society* 3 (November 1999), pp. 908–909.

103. Raoul Bott, "Comments on the Papers in Volume 1," *Raoul Bott Collected Works*, Vol. 1, edited by Robert D. MacPherson (Boston: Birkhäuser, 1994), p. 29.

104. Jackson, "Interview with Raoul Bott."

105. Bott, "Comments," p. 30.

106. Loring Tu (Tufts University), interview with author, March 30, 2012.

107. Ibid.

108. Raoul Bott, "The Stable Homotopy of Classical Groups," *Proceedings of the National Academy of Sciences of the USA* 43 (1957), pp. 933–955; Sir Michael Atiyah, "Raoul Harry Bott," *Biographical Memoirs of Fellows of the Royal Society* 53 (2007), pp. 64–76.

109. Atiyah, "Raoul Harry Bott."

110. Hans Samelson, "Early Days," in *Raoul Bott Collected Papers*, Vol. 1, edited by Robert D. MacPherson (Boston: Birkhäuser, 1994), p. 38.

111. Bott, "Comments," p. 35.

112. Michael J. Hopkins, "Influence of the Periodicity Theorem on Homotopy Theory," in *Raoul Bott Collected Papers*, Vol. 1, edited by Robert D. MacPherson (Boston: Birkhäuser, 1994), p. 52.

113. Allyn Jackson, "Interview with Raoul Bott."

114. Parikh, *The Unreal Life of Oscar Zariski*, p. 113.

115. Raoul Bott, "Comments on Some of the Papers in Volume 2," in *Raoul Bott Collected Works*, Vol. 2, edited by Robert D. MacPherson (Boston: Birkhäuser, 1994), p. xix.

116. Loring Tu (Tufts University), interview with the author, June 14, 2012.

117. Ibid.

118. Raoul Bott, "Comments on Some of the Papers in Volume 4," in *Raoul Bott Collected Papers*, Vol. 4, edited by Robert D. MacPherson (Boston: Birkhäuser, 1995), p. xii.

119. Sir Michael Atiyah, "Raoul Henry Bott," *Biographical Memoirs of Fellows of the Royal Society* 53 (2007), pp. 64–76.

120. Bryan Marquard, "Raoul Bott: Top Explorer of the Math behind Surfaces and Spaces," *Boston Globe*, January 4, 2006.

121. "Bott and Serre Share 2000 Wolf Prize," *Notices of the American Mathematical Society* 47 (May 2000), p. 572.

122. "Bit of Sever's Ceiling Interrupts Math Class," *Harvard Crimson,* November 3, 1966.

123. Alexandra C. Bell, "Math Professor Dies at Age 82," *Harvard Crimson,* January 13, 2006.

124. Lawrence Conlon, in "Raoul Bott as We Knew Him," in *A Celebration of the Mathematical Legacy of Raoul Bott,* edited by P. Robert Kotiuga (Providence, R.I.: American Mathematical Society), pp. 43–50.

125. Barry Mazur (Harvard University), e-mail message to the author, July 31, 2012.

126. Barry Mazur, in "Raoul Bott as We Knew Him," *A Celebration of the Mathematical Legacy of Raoul Bott,* edited by P. Robert Kotiuga (Providence, R.I.: American Mathematical Society), pp. 43–50.

127. Raoul Bott, "Remarks upon Receiving the Steele Career Prize at the Columbus Meeting of the A.M.S.," *Raoul Bott Collected Papers,* Vol. 4, edited by Robert D. MacPherson (Boston: Birkhäuser, 1995), p. 482.

128. Ibid.

Epilogue

1. Benedict Gross (Harvard University), interview with the author, March 28, 2012.

2. Barry Mazur, "Modular Curves and the Eisenstein Ideal," *Publications Mathématiques (IHES)* **47**(1978), pp. 33–186.

3. Barry Mazur (Harvard University), interview with the author, March 28, 2012.

4. Joseph Silverman (Brown University), interview with the author, April 3, 2012.

5. Mazur, interview.

6. John Tate (Harvard University), interview with the author, April 4, 2012.

7. Joseph Silverman (Brown University), e-mail message to the author, April 2, 2012.

8. Tate, interview.

9. Martin Raussen and Christian Skau, "Interview with Abel Laureate John Tate," *Notices of the American Mathematical Society* 58 (March 2011), pp. 444–452.

10. John Tate (Harvard University), interview with the author, December 30, 2009.

11. Tate, interview, April 4, 2012.

12. Mazur, interview.

13. Tate, interview, April 4, 2012.

14. Ibid.

15. Shlomo Sternberg (Harvard University), interview with the author, March 28, 2012.

16. Pat Harrison, "Mathematician Sophie Morel," *Harvard Gazette*, April 16, 2010.

17. Barry Mazur, "Steele Prize for a Seminal Contribution to Research," *Notices of the American Mathematical Society* **47** (April 2000), p. 479.

INDEX

Abelian group, 190
Abelian integrals, 40
Abhyankar, Shreeram, 181–182
Abstract algebra, 25–26, 108
Abstract group, 161
Acta Mathematica, 61, 124
Adams, John, 17–18
Aesthetic Measure (George Birkhoff), 77
Aesthetic theory, 77–78
Agassiz, Louis, 21–23
Ahlfors, Lars, 116; childhood of, 117; education of, 117–118; Denjoy conjecture and, 119–121; publications of, 121, 125, 129–131; teaching by, 121; Nevanlinna theory and, 121–123, 125; covering surfaces and, 124–125; accolades of, 125, 138–139; World War II and, 125–127; full professorship at Harvard, 127; extremal length and, 128–129; quasiconformal mapping and, 129–131; Bers and, 132–133, 136–137; retirement of, 137; alcohol and, 137–138; parties of, 137–138; assault on, 138; death of, 138
Ahlfors-Bers Colloquia, 136–137
Ahlfors finiteness theorem, 136
Aiken, Howard, 143
Aircraft, 142
Air Force, 142
Alaska, 20
Alcohol, 4, 137–138

Algebra: development of, 3, 24; teaching of, 3; Benjamin Peirce work in, 23–27; abstract, 25–26, 108; universal, 108
Algebraic geometry, 168–169, 172, 176, 185
Algebraic Surfaces (Zariski), 171
Algebraic topology, 97, 100, 103, 105, 113
American Ephemeris and Nautical Almanac, 19
American Journal of Mathematics, 27, 33
American Mathematical Monthly, 57
American Mathematical Society, 67
Analysis, 89, 93; harmonic, 46; complex, 116, 129, 139; crypt-, 144–145, 147, 149. *See also* Complex analysis
Annals of Mathematics, 52
Annulus, 62–63, 128
Antiaircraft shells, 142
Anti-Semitism, 80–85, 171
Appel, Kenneth, 79, 100
Applied mathematics, 142–143
Applied Mathematics Group, 110, 142
Arbitrary ground field, 174
Area-length method, 120, 128
Arithmetic, modular, 152
Artificial intelligence, 145
Artin, Emil, 186
Artin, Michael, 186
Artwork, 77

241

Index

"Aryan physics," 83
Astronomy, 16–18
Asymptotic values, 119–120
Atiyah-Bott-Berline-Vergne formula, 201
Atomic bomb, 143–144
Aubin, David, 69
Average, 71–73

Bache, Alexander, 20–22
Ballistics, 142
Bers, Lipman, 132–133, 136–137
Beurling, Arne, 120–121, 127, 131
Birch and Swinnerton-Dyer conjecture, 205
Birkhoff, Garrett: contributions of, 108; publications of, 109; World War II and, 142–143
Birkhoff, George: on Benjamin Peirce, 26; career of, 54–55; skills of, 55; mathematical accomplishments of, 56, 60, 62–63; early life of, 56–57; education of, 57–58; ambitions of, 58; celestial mechanics and, 59, 69; publications of, 62, 64, 67, 69, 75–77; four-color conjecture and, 64–66, 79; minimax principle and, 68; theorem, 68, 108–109; ergodic hypothesis and, 70–74; administrative duties of, 78; gravitational theory and, 78–79; death of, 79; eulogies for, 80; anti-Semitism of, 80–84; legacy of, 86; students of, 86
Black hole, 68–69
Bletchley Park, 145
Bôcher, Maxime, 32, 37; education of, 38–39; publications by, 41, 45–46; potential function and, 44–45; discontinuous functions and, 47; mathematical accomplishments of, 47–48; mathematical style of, 49; teaching by, 50–52; personality of, 52; death of, 53; legacy of, 54
Bôcher prize, 67
Bôcher's theorem, 46
Boltzmann, Ludwig, 70
Bolza, Oskar, 36

Bombes, 145
Born, Max, 158
Born's rule, 158–159
Bott, Raoul, 194; Morse and, 95, 97; electrical network theory and, 195; four-color problem and, 195–196; mathematical interests of, 197; accomplishments of, 198–201; publications by, 199; teaching by, 202; legacy of, 202–203; Harvard math department building and, 207–208
Bott-Duffin theorem, 195
Bowditch, Nathaniel: education of, 8; celestial mechanics and, 8–9; Benjamin Peirce encouraged by, 8–10
Brauer, Richard, 188; teaching by, 189; finite groups and, 189–192; publications of, 192; legacy of, 193–194
Bridge conditions, 153–154
Britain, 145

Cajori, Florian, 3
Calculus: at Harvard, 4; of variations, 88
Calculus of Variations in the Large, The (Morse), 88
Castelnuovo, Guido, 168–171
Category theory, 111–112
Celestial mechanics: Bowditch and, 8–9; George Birkhoff and, 59, 69
Chromatic polynomial, 67
Ciphers, 144
Codebreaking, 144, 146, 149
Cohomology, 104–105, 110; étale, 186; equivariant, 200–201; Tate, 206
Cole, Frank, 38–39
Collaboration, 132–133, 136–137, 159–160, 201
Combinatorics, 99
Comets, 17
Commutative group, 190
Complex analysis, 116, 129, 139
Complex Analysis (Ahlfors), 129
Complex function theory, 116
Computers, 143
Cone, 180

Conformal mapping, 119–120, 128–129
Conjecture: Poincaré, 43, 94; Denjoy's, 118–121; measure zero, 136; Mackey, 157, 159–160; Birch and Swinnerton-Dyer, 205; Sato-Tate, 206. *See also* Four-color conjecture
Connectedness theorem, 176
Continuous functions, 41
Coolidge, Julian, 10–11, 37
CORAL, 146
Court case, 22
Covering surfaces, 124–125
Critical point theory, 90
Cryptanalysis, 144–145, 147, 149
Cubes, 157
Curve, 179; Jordan, 43; stable, 185; elliptical, 204–205; Tate, 206
Cusp, 179

Degeneracy, 174
Deligne, Pierre, 185–186
Deligne-Mumford method, 185
Denjoy's conjecture, 118–121
Deutsche Mathematik, 132
Dieudonné, Jean, 164
Differential equations, 46, 59, 89
Discontinuous functions, 41–42, 46–47
Discrete mathematics, 149
Dogfights, 142
Doob, Joseph, 49, 78
Doughnut, 90–91, 104
Duffin, Richard, 195
Dunster, Henry, 2–3
Dynamical systems, 67
Dynamical Systems (George Birkhoff), 69
Dynamics, 59, 69

Eaton, Nathaniel, 1–2
Education: of Bowditch, 8; of Bôcher, 38–39; of Osgood, 38–39; of George Birkhoff, 57–58; of Ahlfors, 117–118; of Gleason, 164; mathematical, 164
Eilenberg, Samuel, 109–113
Eilenberg-Mac Lane spaces, 112–113
Einstein, Albert, 68–69, 78–79, 83

Eisenbud, David, 185
Electrical network theory, 195
Elements (Euclid), 11
Eliot, Charles, 33–34
Ellipse, 59–60
Elliptical curves, 204–205
Elliptical equations, 204
Elliptical integrals, 204
Embedding, 100
Émigrés, 81, 166
ENIAC, 143
Enigma machine, 145–146
Enormous theorem, 192
Enriques, Federigo, 168, 170, 172
Equations: differential, 46, 59, 89; linear differential, 46; polynomial, 169; elliptical, 204
Equivariant cohomology, 200–201
Ergodic hypothesis, 70–74, 76
Erlangen, 39–40
Étale cohomology, 186
Euclid, 11
Euclidean geometry, 89
Euclidean group, 152–153
Euclidean space, 162
Euler, Leonhard, 12
Euler characteristic, 123
Europe, 35–36
Exeter, 107
Extremal length, 128–129

Farrar, John, 5
Feit-Thompson theorem, 192
Fermat, Pierre de, 57
Fermat's last theorem, 57, 205
Fiber bundles, 101–103
Fields Medal, 2, 183, 186
Fifth problem, 150–152, 155–157
Finite field, 174
Finite groups, 189–192
Finiteness theorem, 136
Finland, 126
Fixed-point theorem, 200
Four-color conjecture: George Birkhoff and, 64–66, 79; Whitney and, 99–100; Bott and, 195–196
Fractals, 43–44
Friedrichs, Kurt, 87

Index

Functions: continuous, 41; discontinuous, 41–42, 46–47; noncontinuous, 41–42; potential, 44–45; Green's, 46; order of, 119; meromorphic, 123; rational, 123–124
Functions of a Complex Variable (Osgood), 54
Functions of Real Variables (Osgood), 54
Function theory, 40, 44
Fundamentals of Abstract Analysis (Gleason), 164

Galle, Johann, 17–18
Galois, Évariste, 169, 190
Galois group, 169
Galois theory, 169–170
Gauss, Carl, 16, 156
Gauss-Bonnet theorem, 125
Generalized geodesic, 68
Geodesics, 68, 88–89
Geodesy, 20
Geometric Integration Theory (Whitney), 105
Geometric invariant theory, 184
Geometric topology, 121
Geometry, 97; Euclidean, 89; algebraic, 168–169, 172, 176, 185
Germany: advanced mathematical study in, 11; as mathematical world leader, 27; graduate training in, 37, 39–40; publication in, 44–45; affectations from, 50; tank production of, 147
Gibbs, J. Willard, 47
Gibbs phenomenon, 47
Gleason, Andrew, 144; Enigma machine and, 146; World War II and, 146–147; career of, 147–148; publications of, 148, 164; mathematical interests of, 149–151, 163–164; Hilbert's fifth problem and, 150–152, 155–157; Mackey and, 159–160; death of, 163; education and, 164
Gleason's theorem, 157
Göttingen, 39–40, 106–107
Graph theory, 99–100
Gravitation, 68, 78–79
Greeks, 156–157
Green's function, 46
Greenwood, Isaac, 4
Grothendieck, Alexander, 177, 185–186
Grötzsch, Herbert, 130
Ground field, 174
Group: Kleinian, 38, 134–136; Lie, 151–153, 198; Euclidean, 152–153; sub-, 153–154; abstract, 161; virtual, 163; Galois, 169; finite, 189–192; abelian, 190; commutative, 190; quotient, 190; simple, 190; monster, 193; special orthogonal, 193; sporadic, 193; homotopy, 197; special unitary, 199; Lubin-Tate, 206; Tate-Shafarevich, 206
Group operations, 162
Group representations, 162
Group theory, 108
Guthrie, Francis, 65

Haken, Wolfgang, 79, 100
Hamilton, William, 24–25
Harmonic analysis, 46
Harris, Joe, 185
Harvard, John, 1
Harvard College Observatory, 16
Harvard Mark I, 143
Harvard Science Center, 208
Harvard University: accomplishments of, 1–2; early history of, 1–2; early curriculum at, 3; calculus at, 4; Benjamin Peirce at, 11; as teaching college, 16, 33; mathematical research at, 32, 41; Ph.D. programs at, 33; Osgood at, 53–54; maturation of, 64; anti-Semitism of, 81; Ahlfors at, 127; World War II and, 141; Zariski at, 177; math department building at, 207–208; Tate at, 207–208; continuity at, 209
Hawkes, Herbert, 28
Heisenberg uncertainty principle, 157
Henry, Joseph, 21–22
Hilbert, David, 151
Hilbert's fifth problem, 150–152, 155–157
Hill, Thomas, 16
Hironaka, Heisuke, 182; on Zariski, 173–174; singularities and, 179–183

Index

Hitler, Adolf, 107
Hodge-Tate decompositions, 206
Holmes, Oliver Wendell, 30
Homeomorphism, 43
Homology, 104, 109. *See also* Cohomology
Homotopy equivalence, 113
Homotopy groups, 197
Homotopy principle, 101
Honda-Tate theorem, 206
H principle, 101
Hydrogen bomb, 143
Hyperplanes, 126

Ideality in the Physical Sciences (Benjamin Peirce), 29
Idempotent, 25–26
Imaginary numbers, 25
Immersion, 100
Immersion theory, 101
Index of stability, 90
Index theorem, 62
Induced representations, 163
Institute for Advanced Study, 95, 172
Invariant, 102, 197
Invariant point, 63
Island analogy, 92–93
Italy, 168–171

Jews, 80–84, 106, 166
Johns Hopkins University, 34–35
Jordan, Camille, 43
Jordan curve, 43

Kelvin, Lord, 20
Kempe, Alfred, 48, 65
Kepler, Johannes, 60
Kleiman, Steven, 177–178
Klein, Felix, 38–39, 44–45, 134
Kleinian groups, 38, 134–136
Knot theory, 205
Krantz, Steven, 129
K-theory, 199
Kummer, Ernst, 67

Langer, Rudolph, 80
Laplace, Pierre-Simon, 8–9, 46
Lavrentev, Mikhail, 130

Lawvere, William, 112
Lazzaroni, 22
Lebesgue, Henri, 41–42
Lectures: by Benjamin Peirce, 13–14; by Morse, 92
Lefschetz, Solomon, 83, 200
Lefschetz fixed-point theorem, 200
Lehto, Olli, 138, 140
Le Verrier, Joseph, 17–18
Libraries, 1
Lie, Sophus, 152
Lie group, 151–153, 198
Lindelöf, Ernst, 117–118
Linear Associative Algebra (Benjamin Peirce), 23, 26–27
Linear differential equations, 46
Lipman, Joseph, 178
Logic, 28, 107
London Mathematical Society, 27
Lubin-Tate group, 206

Machine gun, 142
Mackey, George: physics and, 158; Gleason and, 159–160; discipline of, 160–161; mathematical interests of, 161; representation theory and, 161–163; death of, 163
Mackey conjecture, 157, 159–160
Mackey machine, 163
Mac Lane, Saunders, 81, 83, 106; education of, 106–107; publications of, 109, 112; Eilenberg and, 110–113; category theory and, 111–112; teaching by, 112; at University of Chicago, 113–114; World War II and, 142
Mandelbrot, Benoit, 43–44
Manhattan Project, 141, 143–144
Manifold, 100, 151–153
Map, 42
Mapping: Riemann, 42–43; conformal, 119–120, 128–129; quasiconformal, 129–131; Morrey-Ahlfors-Bers measurable, 133
Mapping theorem, 134
Maschke, Heinrich, 36
Mathematical journals, 11. *See also specific journal*

Mathematical logic, 107
Mathematical Miscellany, 11
Mathematics: in seventeenth century, 3; advanced study in, 11; Ph.D. programs in, 11; America as backwater of, 12; as science of quantity, 24; freedom of, 24–25; rules of, 25; pure, 26, 143; definition of, 28, 48; physical phenomena described by, 29; research in, 32; at University of Chicago, 36; philosophical questions of, 48; approaches to, 49; as competition, 95–96; abstraction of, 97; during retirement, 137; applied, 142–143; discrete, 149; education, 164; as river, 209
Mathematics Genealogy Project, 86
Matrices, 161–162
Maxwell, James Clerk, 20, 70
Mazur, Barry, 205–206
McCleary, John, 142
Mean ergodic theorem, 74
Measure zero conjecture, 136
Mécanique Céleste (Laplace), 8
Meromorphic functions, 123
Method: area-length, 120, 128; Monte Carlo, 143; Deligne-Mumford, 185
Minimax principle, 68, 94
Möbius transformations, 134
Modular arithmetic, 152
Moduli space, 184–185
Monster group, 193
Monte Carlo method, 143
Montgomery, Deane, 154–156
Moore, Eliakim, 36, 58–59
Morrey, Charles, 87, 99, 133
Morrey-Ahlfors-Bers measurable mapping theorem, 133
Morse, Marston, 68, 80, 88; lectures of, 92; academic affiliation of, 95; Bott and, 95, 97; energy of, 95–96; publications of, 96; accomplishments of, 96–97
Morse-Bott theory, 196
Morse relations, 93
Morse theory, 68, 88–92, 94, 96, 196–197

Mountaineering, 98
Mumford, David, 176, 180, 183–185
Music, 77

National Academy of Sciences, 26–27
Nationalism, 80, 84
National Research Council, 99
Navy, 142, 144, 147
Nazis, 83–84, 106–107, 131–132
Neptune, 17–18
Néron-Tate height, 206
Nevanlinna, Rolf, 117–119, 121–122
Nevanlinna theory, 121–123, 125
Newcomb, Simon, 33
New Methods of Celestial Mechanics (Poincaré), 59
Newton, Isaac, 60
New York Mathematical Diary, 11
Nilfactorial, 26
Nilpotent, 25–26
Nim, 149
Nineteenth century, 11
Noether, Emmy, 106, 108, 172–173, 189
Nonabelian group, 191–192
Noncontinuous functions, 41–42
Normal form, 59–60, 67
No small subgroups, 153–154
Number theory, 204–206

Observatory, 16
Osgood, William Fogg, 32, 37; education of, 38–39; function theory and, 40, 44; mathematical accomplishments of, 40, 42–44; publications by, 41–42; mathematical style of, 49; teaching by, 49–51; German affectations of, 50; personality of, 52; scandal of, 53; at Harvard, 53–54; legacy of, 54
Osserman, Robert, 139

p-adic solenoid, 110
Palais, Richard, 153–155
Path spaces, 99
Patriotism, 19
Peacock, George, 24

Pearl Harbor, 141
Peirce, Benjamin, 6–7; perfect numbers and, 7, 11–12; Bowditch encouragement of, 8–10; at Harvard, 11; textbooks by, 13; lectures by, 13–14; teaching by, 14–15; research emphasis of, 16; astronomy and, 18; as patriot, 19; publications of, 19, 27; Saturn and, 19–20; spiritualism-busting activities of, 21–22; in court, 22; mathematical accomplishments of, 23; algebra and, 23–27; George Birkhoff on, 26; death of, 29, 32–33; religiosity of, 29–30; eulogies of, 30–31
Peirce, Charles, 21, 33
Perfect numbers, 7, 11–12
Periodicity theorem, 199
Phase space, 70
Ph.D. programs: in nineteenth century America, 11; at Harvard, 33
Physical phenomena, 29
Physics, 79; "Aryan," 83; Mackey and, 158; group representations and, 162
Picard's theorem, 122
Pless, Vera, 148
Poetry, 77–78
Poincaré, Henri, 43, 56, 59–62, 69, 71, 134, 136
Poincaré conjecture, 43, 94
Pointwise ergodic theorem, 73
Polygons, 77, 156
Polynomial equations, 169
Potential function, 44–45
Principle: minimax, 68, 94; H, 101; homotopy, 101; Heisenberg uncertainty, 157
Proceedings of the National Academy of Sciences, 75
Publications: by Benjamin Peirce, 19, 27; by Bôcher, 41, 45–46; by Osgood, 41–42; in Germany, 44–45; by George Birkhoff, 62, 64, 67, 69, 75–77; by Morse, 96; by Garrett Birkhoff, 109; by Mac Lane, 109, 112; by Ahlfors, 121, 125, 129–131; by Gleason, 148, 164; by Brauer, 192; by Bott, 199; by Tate, 206–207

Pure mathematics, 26, 143
Pythagoras, 77

Quantum mechanics, 162
Quantum theory, 87
Quasiconformal mapping, 129–131
Quasi-ergodic hypothesis, 75–76
Quaternions, 24–25
Quincy, Josiah, 12
Quotient group, 190

Racism, 80–85
Radar, 141
Radicals, 169–170
Ramsey, Frank, 149
Ramsey number, 150
Ramsey theory, 149–150
Rational function, 123–124
Recurrence theorem, 71
Refugees, 81, 84–85, 109, 166
Regular polygons, 156
Religion, 29–30
Representation theory, 161–163
Research: Benjamin Peirce emphasis on, 16; at Harvard, 32, 41; universities focused on, 34; European model of, 35–36
Riemann, Bernhard, 39, 116
Riemann hypothesis, 157
Riemann mapping theorem, 42–43
Riemann sphere, 133, 134
Riemann surfaces, 39, 116, 123, 131, 133–136
Roller Coaster, 180
Rosenlicht, Maxwell, 178
Rotation, 152
Russell, Bertrand, 48
Russian Revolution, 167–168

Samelson, Hans, 199
Sato-Tate conjecture, 206
Saturn, 19–20
Scheffers, Georg, 28
Schmidt, Erhard, 188
Schoenflies problem, 205
Schur, Issai, 188
Schwarzschild, Karl, 68–69

Index

Schwarzschild black hole, 68
Science, 53
Seahorse, 145–146
Séance, 21
Serre-Tate deformation theory, 206
Servais, Cl., 12
Seventeenth century, 3
Simple group, 190
Singularities, 178–183
Smale, Stephen, 89, 94
Society of Fellows, 108
Solenoid, 110
Solitude, 98
Space average, 71–73
Special orthogonal group, 193
Special unitary group, 199
Spencer, Joel, 164
Sphere, 123, 133, 134
Sphere bundles, 101–102
Sphere spaces, 101
Spiritualism, 21–22
Sporadic groups, 193
Stable curves, 185
Statistical mechanics, 73
Statistics, 73, 143, 147
Sternberg, Shlomo, 209
Stiefel-Whitney class, 102–103
Stone, Marshall, 86–87, 113–114
Stone Age, 87
Stone-von Neumann-Mackey theorem, 158
Stone-von Neumann theorem, 157
String theory, 139
Study, Eduard, 27
Subgroup, 153–154
Surveying, 20
Survey of Modern Algebra (Garrett Birkhoff and Mac Lane), 109
Sweden, 126–127
Switzerland, 127
Sylvester, James, 12, 34–36
Symmetry transformation, 152

Tangent bundle, 103
Tate, John, 204; mathematical interests of, 205–206; terms named after, 206; publications of, 206–207; Harvard math building and, 207–208; accolades of, 208–209
Tate cohomology, 206
Tate curve, 206
Tate cycle, 206
Tate module, 206
Tate motive, 206
Tate's algorithm, 207
Tate-Shafarevich group, 206
Tate trace, 206
Taubes, Clifford, 201–202
Teaching: of algebra, 3; by Benjamin Peirce, 14–15; Harvard emphasis on, 16, 33; by Osgood, 49–51; by Bôcher, 50–52; by Mac Lane, 112; by Ahlfors, 121; by Zariski, 175; by Brauer, 189; by Bott, 202
Teichmüller, Oswald, 131–132
Teissier, Bernard, 178
Tennyson, Alfred Lord, 77
Textbooks, 13
Theorem: Riemann mapping, 42–43; Bôcher's, 46; Fermat's last, 57, 205; index, 62; George Birkhoff's, 68, 108–109; recurrence, 71; pointwise ergodic, 73; mean ergodic, 74; uniqueness, 87; Picard's, 122; Gauss-Bonnet, 125; Morrey-Ahlfors-Bers measurable mapping, 133; mapping, 134; uniformization, 134; Ahlfors finiteness, 136; finiteness, 136; Gleason's, 157; Stone-von Neumann, 157; connectedness, 176; main, 176; enormous, 192; Feit-Thompson, 192; Bott-Duffin, 195; periodicity, 199; fixed-point, 200; Lefschetz fixed-point, 200; Honda-Tate, 206
Theory: function, 40, 44; Morse, 68, 88–92, 94, 96, 196–197; aesthetic, 77–78; quantum, 87; critical point, 90; graph, 99–100; immersion, 101; group, 108; category, 111–112; complex function, 116; value distribution, 121; Nevanlinna, 121–123, 125; string, 139; Ramsey, 149–150; representation, 161–163; Galois, 169–170; electrical network,

195; Morse-Bott, 196; K-, 199; number, 204–206; knot, 205; Serre-Tate deformation, 206
Thomson, William, 20
Three-body problem, 56, 60–62
Time average, 71–73
Topology, 64, 69, 89–90, 92–93, 109–110, 151, 154; algebraic, 97, 100, 103, 105, 113; Whitney and, 100; geometric, 121
Torpedo, 142
Torus, 104, 196–198
Transactions of the American Mathematical Society, 46, 52
Triangle, 90
Turing, Alan, 145, 147
Two-body problem, 60
Tyler, Harry, 39–40

Ulam, Stanislaw, 82, 143–144
Uniformization theorem, 134
Uniqueness theorem, 87
Unitary representations, 162
United States: as mathematical backwater, 12; as science world leader, 19; mathematical great awakening in, 37; refugee mathematicians in, 81, 84–85
Universal algebra, 108
University of Chicago, 57–58; mathematics at, 36; Mac Lane at, 113–114
U.S. Coast Survey, 20–21, 22, 24, 27

Value distribution theory, 121
Vandiver, Harry, 57, 65, 79
Varadarajan, Veeravalli, 158, 160, 163

Veblen, Oswald, 59, 80
Vector bundle, 103
Virtual groups, 163
von Neumann, John: ergodic hypothesis and, 71–74; Manhattan Project and, 143–144

Walsh, Joseph, 87–88
Weyl, Hermann, 106–107, 195
Whitehead, Alfred North, 108
Whitney, Hassler: solitude and, 98; mathematical interests of, 99; four-color conjecture and, 99–100; topology and, 100; immersion theory and, 101; fiber bundles and, 101–103; death of, 105
Wiener, Norbert, 42, 50, 63, 70, 82–83
Winthrop, John, 4–5
Witten, Edward, 94
Wolf Prize, 138, 187
World War I, 167
World War II, 110; Ahlfors and, 125–127; academia and, 141; Harvard and, 141; Mac Lane and, 142; George Birkhoff and, 142–143; Gleason and, 146–147; refugees from, 166

Zariski, Oscar, 166; childhood of, 167; mathematical interests of, 168–173; Hironaka on, 173–174; teaching by, 175; main theorem and, 176; at Harvard, 177; singularities and, 178–182
Zund, Joseph, 47, 89